IV International Scientific and Practical Conference on Virtual Simulation, Prototyping and Industrial Design (VSPID-2017)

Journal of Physics: Conference Series
Volume 1084

Tambov, Russia
15 - 17 November 2017

Editor:

Sergei Gorlatch

ISBN: 978-1-5108-7512-8
ISSN: 1742-6588

Printed from e-media with permission by:

Curran Associates, Inc.
57 Morehouse Lane
Red Hook, NY 12571

Some format issues inherent in the e-media version may also appear in this print version.

Printed by Curran Associates, Inc. (2018)

For permission requests, please contact the Institute of Physics
at the address below.

Institute of Physics
Dirac House, Temple Back
Bristol BS1 6BE UK

Phone: 44 1 17 929 7481
Fax: 44 1 17 920 0979

techtracking@iop.org

Additional copies of this publication are available from:

Curran Associates, Inc.
57 Morehouse Lane
Red Hook, NY 12571 USA
Phone: 845-758-0400
Fax: 845-758-2633
Email: curran@proceedings.com
Web: www.proceedings.com

Table of contents

Volume 1084

IV International Scientific and Practical Conference "VIRTUAL SIMULATION, PROTOTYPING AND INDUSTRIAL DESIGN 2017"

15-17 November 2017, Tambov, Russian Federation

Accepted papers received: 10 August 2018
Published online: 5 September 2018

Preface

Papers

IOP Conf. Series: Journal of Physics: Conf. Series **1084** (2018) 011001 doi:10.1088/1742-6596/1084/1/011001

IV International Scientific and Practical Conference "VIRTUAL SIMULATION, PROTOTYPING AND INDUSTRIAL DESIGN – 2017" (VSPID-2017) 15-17 November, 2017 Tambov, Russia

The IV International Scientific and Practical Conference "VIRTUAL SIMULATION, PROTOTYPING AND INDUSTRIAL DESIGN – 2017" (VSPID-2017) has gathered together experienced and young researchers to reveal the potential and prospects of methods, algorithms, and approaches which allow creating and applying virtual models and prototypes in design, modernization, and conversion in different manufacturing sectors, as well as in cultural and architectural preservation.

The foremost aim of the VSPID-2017 was to present and discuss scientific trends, issues and achievements in the field of virtual simulation, prototyping, and industrial design. The topics of the VSPID-2017 were: (1) Virtual modeling, visualization and prototyping of technical systems; (2) Virtual modeling, visualization and prototyping of social objects; (3) virtual modeling in geographic information systems of territories management: development and use; (4) design and application of computer simulation models, virtual simulators and machine vision systems; (5) creation of e-learning materials using virtual simulation and prototyping; (6) industrial design; (7) modern information processing; and (8) modeling of processes of teaching and management in education.

VSPID-2017 was jointly organized by Tambov State Technical University, Russian Fund for Basic Research, and Tambov Region's Government (Russia). The Organizing Committee was introduced by leading professors and specialists from Russian Academy of Sciences, Torino Polytechnic Institute (Italy), University of Soka (Japan), Amberg-Weiden University of Applied Sciences (Germany), University of Münster (Germany), Moscow State Automobile and Road Technical University (Russia), Tambov State Technical University (Russia), Moscow State University (Russia) and other institutions.

All participants were invited to present their papers in this Issue and all submitted manuscripts went through the independent peer review process. We are very grateful to all reviewers from Russia, Italy, China, USA, Canada, Iran, Ukraine, Germany, Serbia, Portugal, Croatia, India and Egypt, for their time and highly professional comments. We deeply believe that their reviews gave opportunity to improve the scientific quality of the presented papers which may be useful for academic, scientific and industrial sectors.

We would like to thank all participants for their contribution, sponsors for the financial support, and Program Committee members for their huge efforts in organization and holding of the VSPID-2017 Conference.

Organizing Committee:
Dr. Vladimir Nemtinov,
Tambov State Technical University,
Tambov, Russia

Editor-in-chief of the Issue:
Prof. Sergei Gorlatch,
Westfälische Wilhelms-Universität,
Münster, Germany

IOP Conf. Series: Journal of Physics: Conf. Series **1084** (2018) 011002 doi:10.1088/1742-6596/1084/1/011002

Peer review statement

All papers published in this volume of *Journal of Physics: Conference Series* have been peer reviewed through processes administered by the proceedings Editors. Reviews were conducted by expert referees to the professional and scientific standards expected of a proceedings journal published by IOP Publishing.

An impact of non-Newtonian blood viscosity on hemodynamics in a patient-specific model of a cerebral aneurysm

D Liepsch[1], S Sindeev[2] and S Frolov[2,*]

[1]Munich University of Applied Sciences, Department of Building Services Engineering and Print and Media Technology, 80335 Munich, Germany
[2]Tambov State Technical University, Biomedical Engineering Department, 392000 Tambov, Russia

[*]E-mail: sergej.frolov@gmail.com

Abstract. Despite of the recent achievements in the research of blood rheology, the effect of non-Newtonian blood viscosity on hemodynamic parameters in cerebral aneurysms is not clarified. The purpose of this study is to investigate the role of different rheological models for prediction of hemodynamic parameters in the patient-specific model of a saccular aneurysm. Transient flow simulations were conducted using OpenFOAM CFD Toolbox. The simulations employed different viscosity models: Newtonian, Power Law, Bird-Carreau, Casson and Local viscosity model. The similar flow patterns were observed both for Newtonian and non-Newtonian fluids in the patient-specific model of the aneurysm. The average difference between Newtonian and non-Newtonian models was about 12%. However for some specific points in the region of recirculating flow the difference was significantly higher - from 20% to 63%. The non-Newtonian blood behaviour influenced the streamlines distribution in the aneurysm region, producing a larger recirculating zone in the center of the aneurysm. The results showed that the major differences between Newtonian and non-Newtonian fluids were found in regions of low velocities and recirculating zones.

1. Introduction

A cerebral aneurysm is a major cerebrovascular disorder which can lead to subarachnoid hemorrhage. There are many known factors contributing to the intracranial aneurysms growth and rupture like hemodynamic stress, arterial wall degeneration, inflammation processes etc., however pathogenesis involving the formation of a cerebral aneurysm remains unclear. A majority of researchers supports a hemodynamic theory of aneurysm genesis and consider a wall shear stress (WSS) as one of the main factors contributing to development of cerebral aneurysms [1-4]. WSS is a friction force which blood flow exerts on the vessel wall that commonly used to evaluate a hemodynamic stress. WSS depends on viscosity value and velocity gradient near the vessel wall, which should be precisely defined for correct evaluation of the stress. A viscosity of blood non-linearly depends on the velocity gradient; however most of the researchers assume blood as a fluid with constant viscosity (Newtonian fluid), which may lead to overestimation of WSS and velocity gradients in some cases [5-7].

The effect of Newtonian viscosity assumption on hemodynamic forces within cerebral aneurysms was evaluated in the study [8]. CFD analysis was performed both for Newtonian and non-Newtonian viscosity models. Newtonian fluid model resulted in a higher aneurysmal time-averaged WSS

IOP Conf. Series: Journal of Physics: Conf. Series **1084** (2018) 012001 doi:10.1088/1742-6596/1084/1/012001

(TAWSS), specifically in the areas of slow flow. In another study Evju et al. [9] investigated the effect of 4 different viscosity models, 2 different inflow conditions and 2 different outflow conditions in 12 aneurysms of middle cerebral artery. They found a strong correlation between the different viscosity models and boundary conditions. No strong correlations were found between the different WSS metrics and the geometrical metrics. Also Morales et al. investigated the role of blood viscosity on intra-aneurysmal hemodynamics in coiled aneurysm models [10]. Numerical simulations were performed to compare the flow fields for Newtonian and non-Newtonian models of fluid. By comparing two fluid models, it was found that Newtonian fluid overestimated the intra-aneurysmal hemodynamics in coiled models. In the high-viscosity regions the viscosity was elevated more than twice and flow velocity was near zero. However for the untreated simulation the Newtonian model does not significantly affected the overall intra-aneurysmal hemodynamics in comparison with non-Newtonian model. Additionally Gambaruto et al. analyzed the sensitivity of the flow field with respect to geometry of cerebral artery and selected fluid model [11]. They found that the sensitivity to geometry variability was greater, but comparable, to the one of the rheological model.

Despite of the recent achievements in the research of blood rheology, further studies should be done to clarify the effect of a non-Newtonian blood behavior on hemodynamic parameters in the cerebral arteries. In this study we provide an insight of blood dynamics in the patient-specific aneurysm model, using different non-Newtonian models of blood. The influence of viscosity models on intra-aneurysmal hemodynamics is investigated by computational fluid dynamics. Three most widely used non-Newtonian viscosity models are employed to represent blood behavior. Moreover using an available data of blood viscosity, the Local viscosity model is proposed for high-precision simulation of viscosity changes at low shear rates. The numerical results are compared with the data obtained from simulations utilizing the Newtonian fluid assumption.

2. Materials and methods

2.1. *Patient-specific model*

A patient-specific model of a cerebral artery was obtained by CT-angiography from a 54 years old patient with a saccular aneurysm. The raw data was post-processed to construct a 3D STL model of the internal carotid artery including an aneurysm sac and inlet and outlet segments. The model was processed to exclude artificial defects. The final geometry of the cerebral artery with the aneurysm is presented in Figure 1. The aneurysm is characterized by the following dimensions. The diameter of the inlet segment d_{inlet} was 3.2 mm and diameters of the outlet segments d_{outlet} were 3 mm. The aneurysm height a_{height} was 12 mm, the aneurysm width a_{width} was 11 mm, the aneurysm depth a_{depth} was 7 mm and the aneurysm neck a_{neck} was 8.13 mm.

2.2. *Boundary conditions for the aneurysm model*

A realistic velocity curve was applied at the inlet segment of the artery (Figure 2). Average velocity over the cardiac cycle was $\overline{u}$ = 238 mm/s. The applied velocity curve corresponds to flow rate of 7 l/h, which is physiologically relevant for the cerebral artery. For simulations with non-Newtonian fluid models a pulse frequency f of 60 beats per minute (one beat per second) was used. A free outflow boundary condition was imposed at the outlet segments of the artery.

IOP Conf. Series: Journal of Physics: Conf. Series **1084** (2018) 012001 doi:10.1088/1742-6596/1084/1/012001

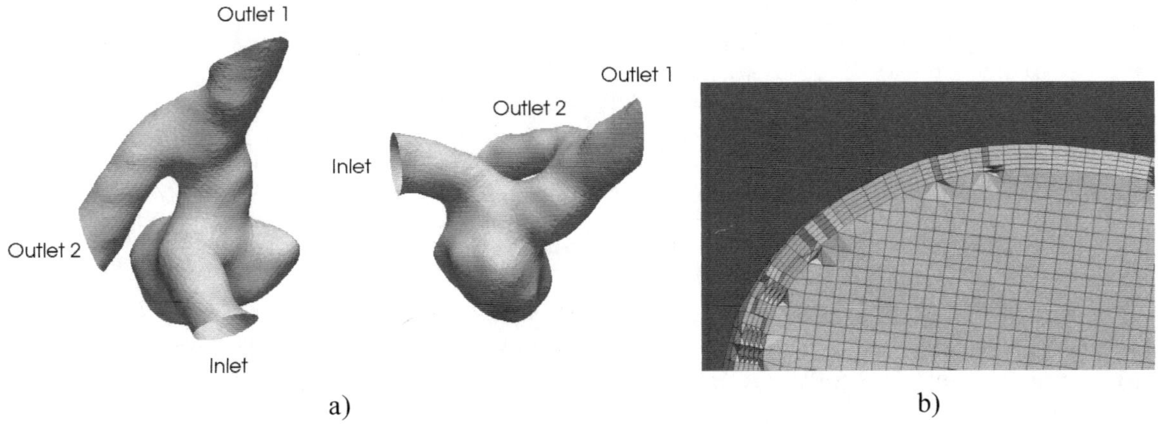

Figure 1. Patient-specific model of the cerebral aneurysm: a) aneurysm sac at different projections; b) near-wall cell layers.

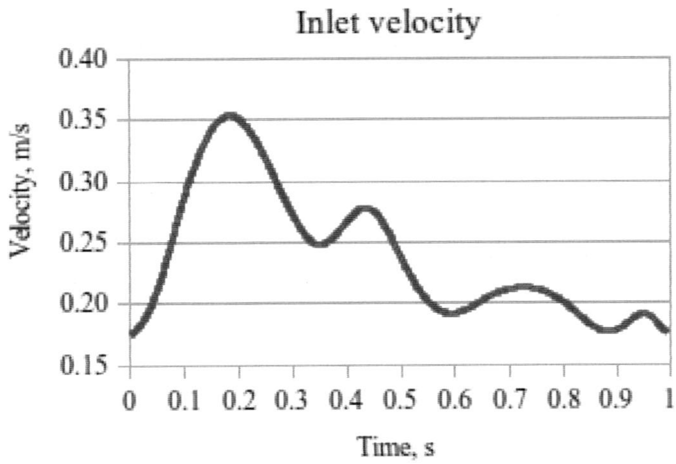

Figure 2. Inlet velocity curve for the studied saccular aneurysm model.

In order to compare the simulation results for Newtonian and non-Newtonian fluid models the dimensionless parameters - Reynolds number Re and Strouhal number St should be the same [12, 13]. To calculate Re for non-Newtonian fluid a representative viscosity value η_{rep} should be found. This value can be obtained using the coefficients k and n of a Power Law model (see Section 2.3) [12, 14]:

$$\eta_{rep} = k\left(\frac{2\pi\bar{u}}{d_{inlet}}\right)^{n-1} \cdot \frac{\pi}{4}\left(\frac{3n+1}{\pi n}\right)^{n} = 3.5 \text{ mPa•s}.$$

Using the representative viscosity value the Reynolds number at the inlet segment of the artery could be calculated as

$$Re = \frac{\bar{u} \cdot d_{inlet} \cdot \rho}{\eta_{rep}} = 228,$$

where blood density ρ is 1050 kg/m^3.

The corresponding representative shear rate γ_{rep} for the non-Newtonian fluid was calculated by formula:

$$\gamma_{rep} = \frac{8 \cdot \bar{u}}{d_{inlet}} \cdot \left(\frac{3n+1}{4n} \right)^{\frac{n}{n-1}} = 467 \text{ s}^{-1}.$$

The Womersley number α could be found using the equation [14]:

$$\alpha = \frac{d_{inlet}}{2} \cdot \sqrt{\frac{2\pi f}{(\eta_{rep} / \rho)}} = 2.1.$$

The Strouhal number St was

$$St = \frac{f \cdot d_{inlet}}{\bar{u}} = 0.01344.$$

For a correct comparison, the Womersley number α as well as Reynolds Re and Strouhal St numbers for Newtonian and non-Newtonian fluids should be the same [12, 13]. Therefore the average inlet velocity $\bar{u}$ and pulse frequency f were adjusted to obtain similar Re and St for Newtonian fluid model. The average velocity for Newtonian fluid model was $\bar{u}$ = 0.254 m/s and frequency f was 64 beats per minute.

2.3. Fluid viscosity

The two types of viscosity models were used to represent blood: Newtonian and non-Newtonian. A Newtonian model is the simplest assumption for modeling of blood rheology. This model considers blood as fluid with constant viscosity:

$$\eta(\gamma) = \eta, \tag{1}$$

where η is dynamic viscosity; γ is shear rate. The viscosity η = 4.1 mPa•s was used to represent the Newtonian viscosity for numerical studies. This value corresponds to kinematic viscosity of ν = 3.905 • 10^{-6} m^2/s.

To represent the non-Newtonian blood behavior three most widely used models were employed: Power Law, Bird-Carreau and Casson.

Power Law model:

$$\eta(\gamma) = k\gamma^{n-1}, \quad \eta_{min} \le \eta \le \eta_{max}, \tag{2}$$

where k is consistency index; n is an index. The viscosity is bounded by minimum η_{min} and maximum η_{max} values respectively. The following parameters for the Power Law model were used in this study: k = 0.0117642; n = 0.8092; η_{min} = 3.5 mPa•s and η_{max} = 14 mPa•s.

Bird-Carreau model:

$$\eta(\gamma) = \eta_{min} + (\eta_{max} - \eta_{min}) \cdot \left[1 + (k\gamma)^2\right]^{(n-1)/2}. \tag{3}$$

The parameters of Bird-Carreau model were: k = 0.6046; n = 0.3742. Viscosities η_{min} and η_{max} were the same as for the Power Law model.

Casson model:

$$\eta(\gamma) = \left(\sqrt{\frac{\tau_y}{\gamma}} + \sqrt{k} \right)^2. \tag{4}$$

where τ_y is yield stress. The parameter values of τ_y = 3.6; k = 4.1 were used to represent blood with by the Casson model.

Additionally a Local viscosity model was proposed:

$$\eta(\gamma) = a_i + b_i(\gamma - \gamma_i) + c_i(\gamma - \gamma_i)^2 + d_i(\gamma - \gamma_i)^3, \quad [\gamma_{i-1}, \gamma], \; i = \{1, 2, \ldots, S\}, \tag{5}$$

where a_i, b_i, c_i and d_i are coefficients of i-th spline; S is a number of splines. The Local viscosity model uses a set of cubic splines to precisely interpolate the measured blood viscosity. For every time moment the viscosity value is computed using the current velocity gradient at the point.

The presented parameter values for non-Newtonian viscosity models where chosen in such a way to minimize an error between viscosity curves for blood and its corresponding rheological model. The least squares method was employed for this purpose. The viscosity curves for selected fluid models (1)-(5) are presented in Figure 3.

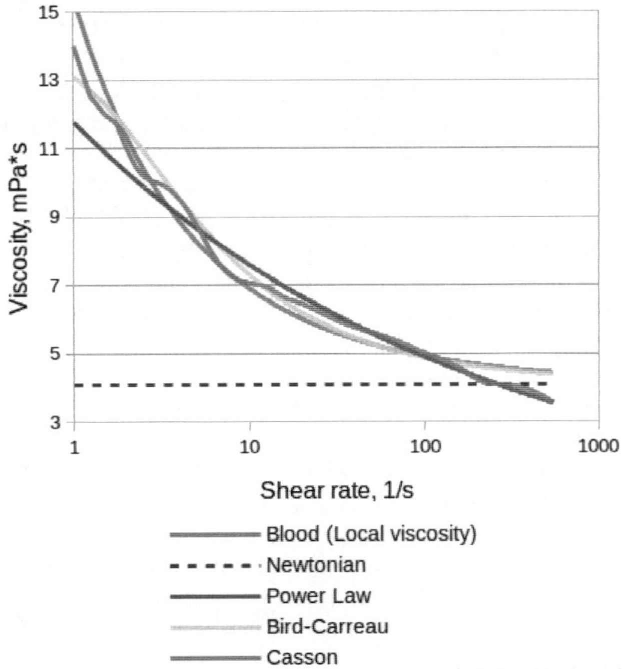

Figure 3. Viscosity over shear rate for the different fluid models.

2.4. Numerical simulation

SnappyHexMesh tool was used to generate a hexahedral computational mesh with maximum edge length of 0.128 mm for numerical studies. The mesh was additionally refined by adding five cell layers (cell height is 0.0375 mm) near the vessel wall for precise computation of the near-wall velocity gradients (see Figure 1b). Mesh independence check was carried out to ensure that results were not affected by selected mesh size. The average y+ value was 0.74. Two million cells were sufficient to capture the flow details in the regions of interest. The same computational mesh was used both for Newtonian and non-Newtonian cases. The flow simulations were based on the three-dimensional incompressible Navier-Stokes and continuity equations:

$$\rho\left(\frac{\partial \mathbf{u}}{\partial t} + \mathbf{u}\cdot\nabla\mathbf{u}\right) = -\nabla\mathbf{P} + \nabla\cdot\tau, \quad (6)$$

$$div\,\mathbf{u} = 0, \quad (7)$$

$$\tau = \eta\gamma\left(\nabla\mathbf{u} + \nabla\mathbf{u}^T\right), \quad (8)$$

where $\mathbf{u}$ is blood velocity; $\mathbf{P}$ is blood pressure; τ is a stress tensor.

The vessel wall was assumed to be rigid and no-slip boundary condition was applied. The simulation was carried out until a converged solution was obtained. The time step for numerical simulations was $\Delta t = 1\times10^{-3}$ s. Open source CFD toolbox OpenFOAM (OpenFOAM Foundation Ltd, London, UK) was used to conduct numerical studies. Second order numerical schemes were used for spatial and temporal discretization. The PIMPLE algorithm was employed for solving the equations (6)-(8).

3. Results

The main hemodynamic parameters in the patient-specific aneurysm model were analyzed to estimate the impact of different rheological models of blood on hemodynamic parameters in the aneurysm sac. For this purpose the region of interest, including the aneurysm sac, was extracted from the dataset. Hemodynamic parameters were analyzed during the systolic peak, which corresponds to the maximal hemodynamic stress. A velocity distribution for the central cross-section of the aneurysm model is presented in Figure 4. The similar flow patterns were found both for Newtonian and non-Newtonian fluids. The high velocity zones were observed near the left and bottom sides of the aneurysm dome, whereas zones of low velocity were found in the center of the aneurysm and in the aneurysm bleb. The maximum velocity value was predicted by Newtonian model; however the relative difference between Newtonian and non-Newtonian models for the maximum velocity was about 5%. To analyze an impact of different fluid models on the velocity distribution a set of 36 key points was employed. Position of these points is presented in Figure 5. Table 1 summarizes the observed velocity values at the key points for used fluid models. According to the Table 1 the Newtonian model showed a higher velocity magnitude in the aneurysm region then other fluid models. The Local viscosity model was used as a reference for comparison with other fluid models, because it precisely represents the measured viscosity of blood. The average difference between Newtonian and Local viscosity model was about 12%. However in some specific points at the region of recirculating flow the difference between the models was significantly higher - from 20% to 63%.

A viscosity distribution for the central cross-section of the aneurysm dome is shown in Figure 6. The high viscosity region was located near the center of the aneurysm sac and in the aneurysm bleb. The maximum viscosity values were observed near the left-bottom region of the aneurysm cross-section. In these regions the viscosity value was 5.27 mPa•s, which is approximately 28.5% higher than for the Newtonian model. However the maximum viscosity value was not reached by any fluid model. The Power Law model showed an underestimation of high viscosity in comparison with the Local viscosity model as well as both Bird-Carreau and Casson models showed an overestimation on 5% and 21.38% respectively.

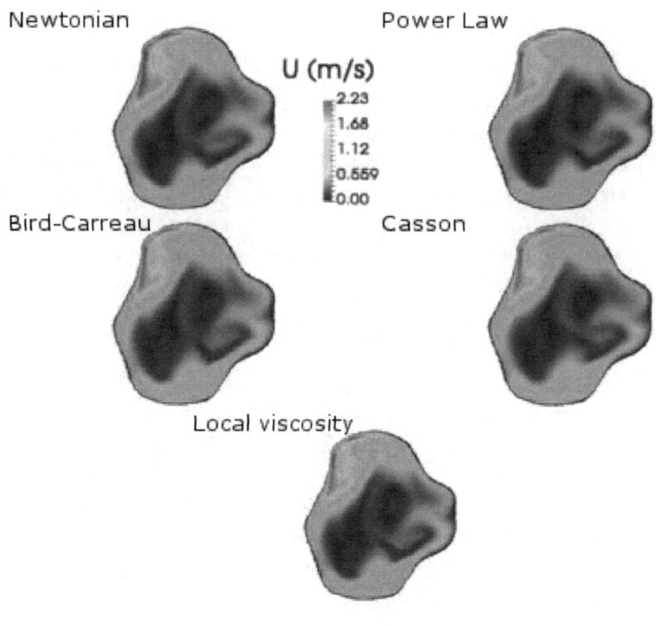

Figure 4. Velocity distribution at the central cross-section.

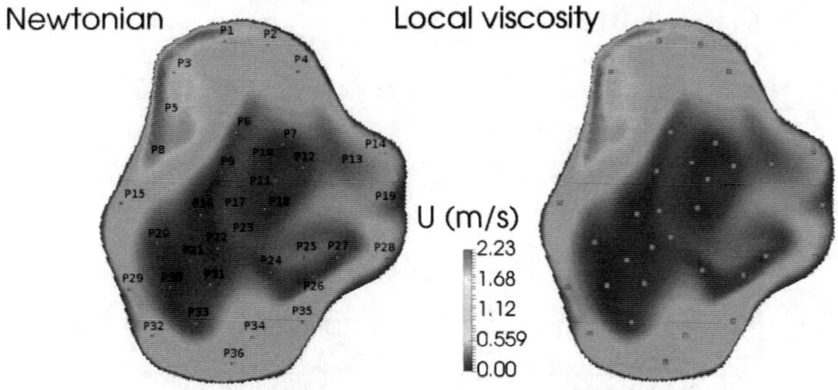

Figure 5. Position of the key points for Newtonian and Local viscosity fluid models.

Table 1. Velocity (m/s) at the key points for the different fluid models.

	Newtonian	Power Law	Bird-Carreau	Casson	Local viscosity	Δ, %
P1	1.44	1.34	1.36	1.40	1.33	7.3
P2	1.21	1.16	1.15	1.14	1.16	4.6
P3	1.60	1.51	1.51	1.51	1.51	5.5
P4	0.95	0.91	0.89	0.90	0.90	5.2
P5	1.94	1.83	1.83	1.84	1.83	5.7
P6	0.21	0.20	0.20	0.19	0.19	6.0
P7	0.10	0.07	0.11	0.13	0.08	23.4
P8	1.94	1.84	1.84	1.83	1.84	5.1
P9	0.12	0.11	0.10	0.10	0.11	6.5
P10	0.15	0.11	0.15	0.14	0.12	22.2
P11	0.14	0.15	0.13	0.11	0.14	1.7
P12	0.13	0.13	0.12	0.10	0.12	6.9
P13	0.28	0.26	0.27	0.27	0.26	8.7
P14	0.54	0.53	0.53	0.51	0.53	2.3
P15	1.09	1.04	1.03	1.02	1.04	4.4
P16	0.08	0.08	0.08	0.11	0.08	0.3
P17	0.23	0.18	0.19	0.13	0.19	17.1
P18	0.11	0.07	0.09	0.15	0.07	35.1
P19	0.21	0.22	0.21	0.18	0.22	9.4
P20	0.13	0.11	0.11	0.14	0.11	16.5
P21	0.09	0.10	0.10	0.12	0.09	0.2
P22	0.06	0.08	0.06	0.05	0.07	12.3
P23	0.25	0.22	0.20	0.11	0.22	11.9
P24	0.09	0.04	0.09	0.34	0.08	7.5
P25	0.38	0.38	0.38	0.29	0.37	2.3
P26	0.10	0.16	0.16	0.14	0.16	63.0
P27	0.13	0.11	0.11	0.14	0.11	14.5
P28	0.47	0.46	0.47	0.40	0.46	2.8
P29	0.60	0.57	0.57	0.58	0.57	4.3
P30	0.03	0.04	0.04	0.04	0.04	20.2
P31	0.12	0.14	0.12	0.11	0.13	7.4
P32	0.72	0.67	0.67	0.68	0.67	7.1
P33	0.13	0.14	0.13	0.13	0.14	9.2
P34	1.04	0.97	0.98	0.98	0.97	6.6
P35	1.10	1.05	1.05	1.04	1.05	4.5
P36	1.29	1.24	1.24	1.22	1.24	4.0

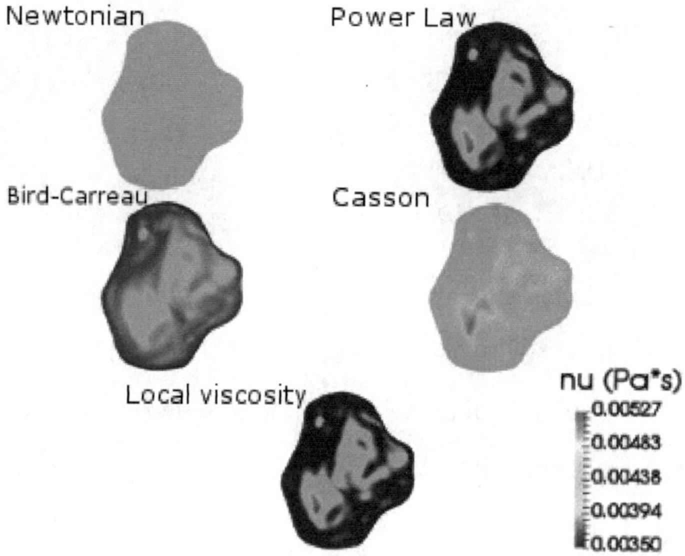

Figure 6. Viscosity distribution for the different fluid models.

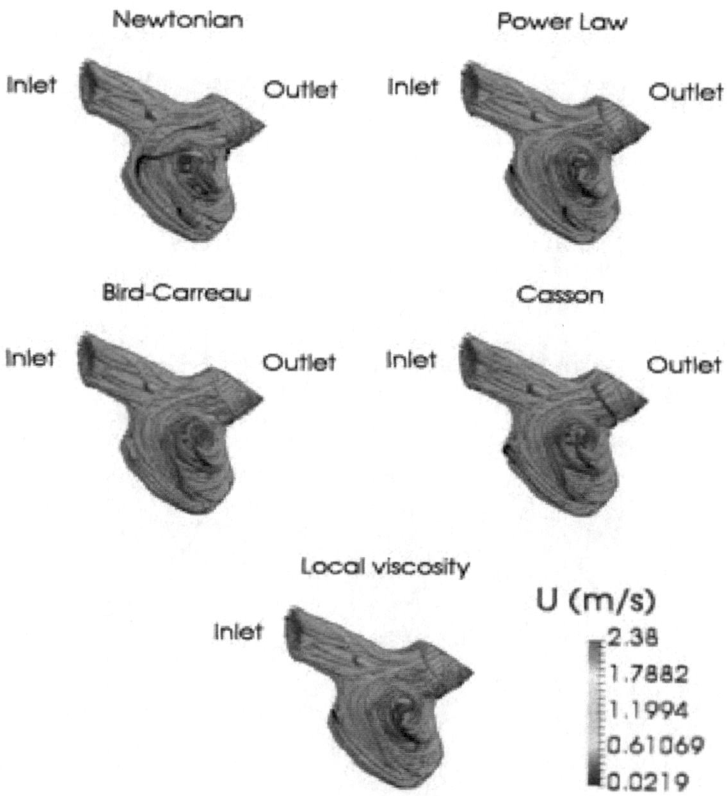

Figure 7. Computed streamlines during the systolic peak.

IOP Conf. Series: Journal of Physics: Conf. Series **1084** (2018) 012001 doi:10.1088/1742-6596/1084/1/012001

Streamlines were computed for all fluid models to visualize a flow pattern in the aneurysm sac. The obtained streamlines are presented in Figure 7. The inlet flow separates on two parts: the first one flow along the parent artery; another one produces a swirl motion in the region of aneurysm. In the center of a zone of recirculation the lowest velocity values were detected. The non-Newtonian blood behavior influenced a streamlines distribution in the aneurysm region producing a larger recirculating zone in the center of the aneurysm.

Using calculated velocity gradients and viscosity values a WSS distribution was obtained for all fluid models. Computed WSS distribution is presented in Figure 8. The general pattern of WSS distribution over the aneurysm surface was similar for Newtonian and non-Newtonian models. For the most regions the WSS value was approximately 20 Pa. However for zones of low WSS, the Newtonian model overestimated the highest value of WSS on 16.5% in comparison with the Local viscosity model.

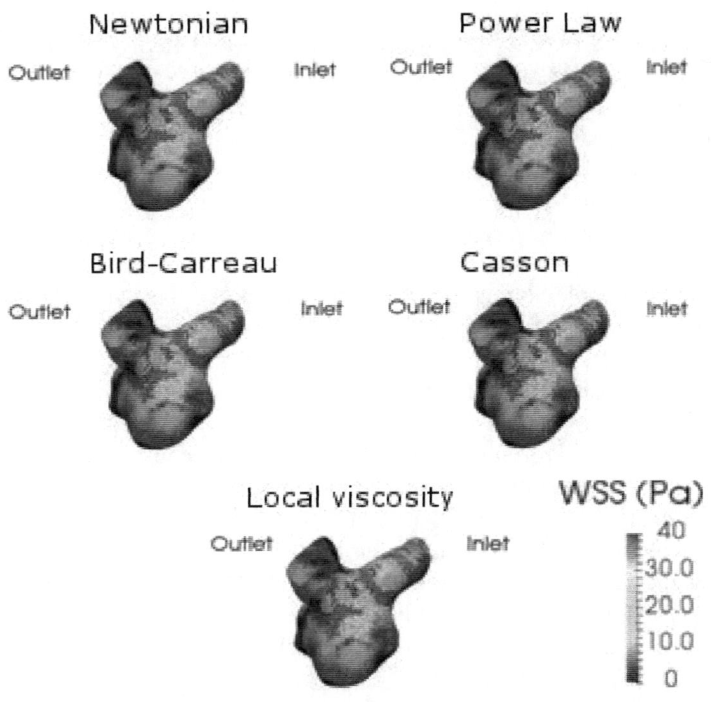

Figure 8. Wall shear stress distribution during the systolic peak.

The results showed that non-Newtonian properties had the significant influence on blood flow characteristics in the studied case, especially at zones of recirculation, and should be considered in the future investigations of cerebral hemodynamics.

4. Discussion

The purpose of the present study was to investigate the effect of non-Newtonian behavior of blood on hemodynamics in a realistic cerebral artery and to evaluate an influence of different viscosity models on the resulting hemodynamic parameters. In this work we presented detailed numerical simulations of hemodynamics in the patient-specific model of the saccular aneurysm using different rheological models: Newtonian, Power Law, Bird-Carreau, Casson and Local viscosity.

Hemodynamic factors such as velocity and pressure gradients, regions of high and low shear stresses play an important role in forming cerebral aneurysms. This is especially important at bifurcations, where the flow is disturbed and secondary flows are created. The Newtonian model of fluid is a common assumption for simulation of blood flow in large arteries, however for small arteries, especially for cerebral arteries with aneurysm; this assumption may not be precise in all cases, since in regions with slow and recirculating flow blood exhibits a non-Newtonian behavior. Moreover it should be noted that the used non-Newtonian fluid models are only the simplifications of blood properties and not necessary depict the blood internal forces. In the present study to overcome the errors of approximation, which are the results of applying a certain non-Newtonian model, the Local viscosity model was used. The Local viscosity model is based on a set of cubic splines which precisely interpolate the measured viscosity data.

Blood flow dynamics was studied under physiologically relevant flow conditions which were adjusted to keep the similar dimensionless parameters between Newtonian and non-Newtonian fluid models. The vessel wall was assumed to be rigid. According to the study [15] the aneurysm wall has a lack of elastic and therefore exhibits a reduced compliance. Hence an assumption of rigid walls is reasonable. Aneurysm geometry has a significant influence on size and location of zones with slow recirculating flow.

In this study, the highest differences between Newtonian and non-Newtonian fluid models were observed especially at these zones, which is in agreement with the recent study by Carty et al. [16]. Furthermore, Gibaly et al. [17] found, that despite of the general agreement of flow patterns, at low velocity zones the non-Newtonian fluid model predicted lower shear stresses compared to Newtonian model, which could significantly affect hemodynamic parameters at these specific regions. In our case the average difference between Newtonian and Local viscosity models was about 12%; however, at some specific points inside zones of recirculating flow the difference was from 20% to 63%. It also states that non-Newtonian model should be used for flow studies in the cerebral aneurysms. Gambaruto et al. [18] employed a Newtonian model and two generalized non-Newtonian models – Carreau and Cross viscosity model for flow studies in cerebral aneurysm models and found variability of the results between the different viscosity models. In contrast, in our study we showed that even usage of patient-specific viscosity data with standard non-Newtonian models could produce significant differences, especially at the low shear rates zones, compared to the precise blood viscosity curve (Local viscosity model). Moreover, standard non-Newtonian viscosity models could not fit perfectly the viscosity values both at high and low shear rates simultaneously due to predefined form of the curve. Either low or high shear rates values could be fitted well. This is supported by the recent study by Suzuki et al. [19], demonstrating that independent of the aneurysm size, numerical simulations with Newtonian or non-Newtonian fluid model could produce values different from those of the patient-specific viscosity model. Additionally, the results of our study correlate with the recent study by Meng et al. which found that a usage of Newtonian assumption could overestimate the hemodynamic parameters, while underestimate the potential of thrombus formation [20].

The model example, which utilized the patient-specific model of a cerebral artery with aneurysm, demonstrated the importance of choosing a specific non-Newtonian fluid model for study of hemodynamics in a realistic aneurysm. The realistic geometry of the cerebral artery played an important role in the intra-aneurysmal distribution of hemodynamic parameters. The intra-aneurysmal flow showed a complex vortex structure. The increment of viscosity in the center of the vortex and in the aneurysm dome led to reduction of the flow due to the viscous forces. This hemodynamic conditions are very important for thrombus formation and clinically relevant for a decision making process.

Our study extends the recent studies that evaluate effect of non-Newtonian rheological properties of blood on distribution of hemodynamic parameters in regions of bifurcation and aneurysm, however it has some limitations. The obtained numerical results were not validated by high precision measurement technique such as LDA or PIV, e.g. [21]. The number of cases should be extended to derive the statistical implications from the results. Also the assumption of rigid walls was used, which could potentially affect the viscosity distribution. Future studies will address the listed limitations. Furthermore, more investigations should be done to evaluate the viscosity changes in patient-specific models of cerebral aneurysms after a treatment with flow-diverter stents where zones of stagnant flow are created. The further studies include the investigation of viscosity changes in patient-specific models of cerebral aneurysms treated by the flow-diverter.

5. Conclusion

This work presents a numerical study of hemodynamics in a patient-specific model of the cerebral aneurysm using different rheological models of blood. A Newtonian model overestimates the maximum velocity magnitude compared to non-Newtonian models. The study showed that the major differences between Newtonian and non-Newtonian fluids are observed in regions of low velocities and recirculating zones. It was found that for the patient-specific model of the aneurysm the difference in computed velocity magnitude varied from 20% to 63% at some points. It confirms that non-Newtonian model of blood should be used for studies of hemodynamics in the cerebral arteries. Also it is demonstrated that the Local viscosity model allows eliminating approximation errors and it could be used for the cases where the measured data of blood viscosity is available and high precision of estimation of hemodynamic parameters is necessary.

Acknowledgments
This work was supported by Russian Science Foundation (Project №16-15-10327).

References

[1] Jou L, Lee D and Morsi H 2008 *AJNR Am. J. Neuroradiol.* **29**(9) 1761–7
[2] Kulcsar Z, Ugron A, Marosfoi M, Berentei Z, Paal G and Szikora I 2011 *AJNR Am. J. Neuroradiol.* **32**(3) 587–94
[3] Shojima M, Oshima M and Takagi K 2004 *Stroke* **35**(11) 2500–5
[4] Tanoue T, Tateshima S, Villablanca J P, Vinuela F and Tanishita K 2011 *AJNR Am. J. Neuroradiol.* **32**(9) 1732–7
[5] Bernsdorf J and Wang D 2009 *Comput. Math. Appl.* **58**(5) 1024–9
[6] Frolov S V, Sindeev S V, Liepsch D and Balasso A 2016 *Technol. Health. Care.* **24**(3) 317–33
[7] Xiang J, Tremmel M, Kolega J, Levy E I, Natarajan S K and Meng H 2012 *J. Neurointerv. Surg.* **4**(4) 351–7
[8] Hippelheuser J E, Lauric A, Cohen A D and Malek A M 2014 *J. Biomech.* **47**(15) 3695-3703
[9] Evju O, Valen-Sendstad K and Mardal K 2013 *J. Biomech.* **46**(16) 280–8
[10] Morales H, Larrabide I and Geers A 2013 *J. Biomech.* **46**(13) 2158–64
[11] Gambaruto A, Janela J, Moura A and Sequeira A 2011 *Math. Biosci. Eng.* **8**(2), 409–23
[12] Liepsch D 1986 *Biorheology* **23**(4) 395-433
[13] Liepsch D, Sindeev S, Balasso A and Frolov S 2017 *Series on Biomechanics.* **31**(1) 3-11
[14] Balasso A, Bauer J, Liebig T, Dorn F, Zimmer C, Liepsch D and Prothmann S 2014 *Biorheology* **51**(6) 341–54
[15] Humphrey J and Na S 2002 *Ann. Biomed. Eng.* **30**(4) 509–23
[16] Carty G, Chatpun S and Espino DM 2016 *Med. Biol. Eng. Comput.* **36**(3), 396-409
[17] Gibaly A, El-Bassiouny OA, Diaa O, Shehata AI, Hassan T and Saqr KM 2016 *Applied Mechanics and Materials* **819** 366-370
[18] Gambaruto A, Janela J, Moura A and Sequeira A 2013 *Mathematical Biosciences and Engineering: MBE* **10**(3) 649-665
[19] Suzuki T, Takao H, Suzuki T, Suzuki T et al. 2017 *Technol. Health Care* **25**(1) 37-47
[20] Meng H, Tutino V, Xiang J and Saddiqui A 2014 *AJNR Am. J. Neuroradiol.* **35**(7) 1254–62
[21] Frolov SV, Sindeev SV, Liepsch D and Balasso A 2016 *Technol. Health Care* **24**(3) 317-333

IOP Conf. Series: Journal of Physics: Conf. Series **1084** (2018) 012002 doi:10.1088/1742-6596/1084/1/012002

Application of infrared vision system for potato thermal control

A Churikov[1], A Divin[1], A Egorov[1] and L Podestà[2]

[1]Department of Mechatronics and Technological Measurements, Tambov State Technical University, 106 Sovetskaya Street, Tambov 392000, Russian Federation
[2]Department of Astronautics, Electrical and Energetics Engineering, Sapienza University of Rome, via Eudossiana 18, Rome 00184, Italy

[1]E-mail: agdv@yandex.ru

Abstract. The article proposes a solution to the problem of determining the values of potato tubers tissue thermophysical characteristics. The solution of this problem makes it possible to calculate the optimal regime parameters of potato quality active thermal control, which can be used for automatic sorting. In order to solve this problem, we propose a non-contact non-destructive control method based on a pulsed laser heating of a potato tuber flat surface area and subsequent use of time integral characteristics of temperature and heat flow, as well as a measuring device developed on the basis of physical and mathematical models of the method. The method was used by the authors to determine the thermal conductivity, heat capacity and coefficient of thermal diffusivity of different quality potato tissues: both healthy and affected by phyto-diseases. The studies have shown that the thermal conductivity of plant tissues depends on the presence of structural disturbances in them as a result of phyto-diseases. This fact confirms the possibility of using thermal non-destructive control of potatoes tissues provided the correct choice of power and the duration of the thermal effect on the object of control.

1. Introduction

Thermophysical characteristics (TPCs) of potato tissue are responsible for calculating the regime parameters of the potato storage and heat treatment [1]. Meanwhile, potato TPCs are significantly influenced by the physicomechanical, physicochemical and chemical characteristics of potato tissue [2]. Moreover, there is a great difference between heat capacity of healthy plant and heat capacity of defective tissue cause by phyto-diseases or external mechanical effects as been previously revealed in [3]. Thus, thermal control methods can be successfully applied to sort out agricultural products during storage and shipping to the consumer.

Modern methods of plant non-destructive testing are mainly based on the vision systems in the visible spectral range [4] as well as in the near infrared radiation range [5]. Meanwhile, the technical vision systems in the middle range of infrared radiation with the use of thermal imaging cameras have the advantage of image analysis simplification [6]. However, the effectiveness of such control largely depends on the conditions and regime control parameters.

In the process of thermal control, it is important to maintain parameters as the power and duration of thermal stimulation, which must be determined by mathematical modeling methods at TPC values measured during thermophysical experiment. In this connection, the problem of developing a method for non-destructive testing of potato TPCs arises. Meanwhile, plant TPCs measurement is largely

IOP Conf. Series: Journal of Physics: Conf. Series **1084** (2018) 012002 doi:10.1088/1742-6596/1084/1/012002

influenced by specific tissue structure of most vegetables and fruits. Typically, plant tissue is a heterogeneous medium consisting of water, solids and gas (air) pores. Therefore, during the measurement, the following conditions must be fulfilled.

The aim of the paper was to develop a method for non-destructive testing of potato TPCs.

2. Theoretical basis and experimental part

Differently sized and shaped piece of fruits and vegetables were objects of our research. Therefore, it was advisable to take a semi-bounded body as a physical model of the investigated sample. The thermal effect on the surface area of this body could cause a multidimensional thermal process in it. It was possible to restore values of the sample's thermal conductivity and thermal diffusivity by registering the parameters of this process.

2.1. Development of method for non-destructive testing of potato TPCs

The organization of a semi-bounded body heating was carried out through a section of the simplest geometric form. Based on the possibility of practical implementation and a simple mathematical description of the thermal process in a solid, we selected the circle as a sample under study surface section. The heat flux entered through this surface section.

To solve multidimensional heat conduction problems, we used the apparatus of temperature and heat flow integral characteristics. The mathematical models of the relative and absolute methods of non-destructive TPCs testing (thermal conductivity and thermal diffusivity), developed earlier and described in [7-9], assumed thermal action by a constant density of heat flux over time $q(t) \equiv q = const$, due to which the temperature in the sample under study reached a stationary value. In these methods, the main experimental parameter was the time integral characteristic of the surface temperature of a heated sample:

$$S^*(p) = \int_0^\infty e^{-p \cdot t} \cdot S(t)dt, \quad p > 0, \tag{1}$$

where $S(t)$ was measured mean-integral temperature of the heated circle, and p was Laplace transform parameter.

We assumed that in the experiment the investigated body was semi-bounded $0 \le z < \infty$, $0 \le r < \infty$ (Figure 1) and had a constant initial temperature (0 degrees Celsius. We also proposed that the heat flux was time-limited:

$$q(t) = \begin{cases} q(t), t \le t_2, \\ 0, t > t_2; \end{cases}.$$

Thus, in the developed method and apparatus we supposed that the thermal effect was generated by time-limited heat flux.

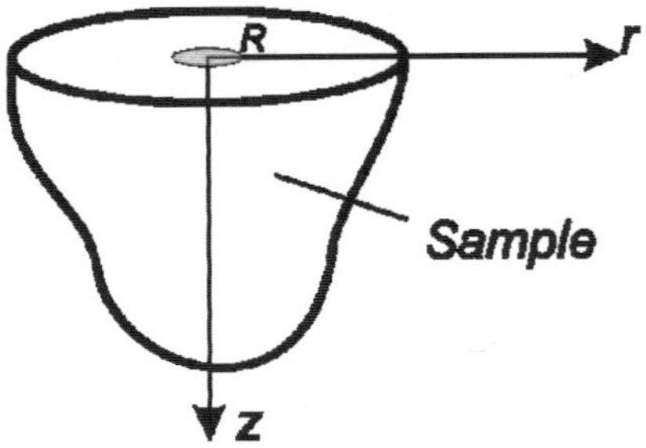

Figure 1. Physical model of the thermal process in the heated body: R – radius of the heated surface area, r - horizontal axis, z - vertical axis.

Under the assumptions, temperature field in the semi-bounded body are described by solving the following axially symmetric boundary value problem:

$$\frac{1}{a}\cdot\frac{\partial U(t,r,z)}{\partial t}=\frac{\partial^2 U(t,r,z)}{\partial r^2}+\frac{1}{r}\cdot\frac{\partial U(t,r,z)}{\partial r}+\frac{\partial^2 U(t,r,z)}{\partial z^2}, \tag{2}$$

$$(t>0,\ 0\le r<\infty,\ 0\le z<\infty);$$

$$U(0,r,z)=0; \tag{3}$$

$$U(t,r,z)=0 \text{ for } r,z\to\infty;\ \left.\frac{\partial U(t,r,z)}{\partial r}\right|_{r=0}=0; \tag{4}$$

$$\lambda\left.\frac{\partial U(t,r,z)}{\partial z}\right|_{z=0}=\begin{cases}-q_1(t,r) \text{ for } 0\le r\le R,\\ 0 \text{ for } r>R;\end{cases}\quad q_1(t,r)=\begin{cases}\varepsilon\cdot q(t) \text{ for } t\le t_2,\\ 0 \text{ for } t>t_2.\end{cases} \tag{5}$$

In equation (5) ε is a coefficient of surface radiation emissivity of the body under investigation.

We apply to the problem (2) - (5) the integral Laplace transform with respect to the time coordinate t, assuming that, when heated, the function $U(t,r,z)$ is continuous and has continuous derivatives with respect to coordinates r and z in the region, $0\le r<\infty$, $0\le z<\infty$. We obtain an equation:

$$\frac{p}{a}\cdot U^*(p,r,z)=\frac{\partial^2 U^*(p,r,z)}{\partial r^2}+\frac{1}{r}\cdot\frac{\partial U^*(p,r,z)}{\partial r}+\frac{\partial^2 U^*(p,r,z)}{\partial z^2}, \tag{6}$$

with boundary conditions

$$U^*(p,r,z)=0 \text{ for } r,z\to\infty;$$

$$\left.\frac{\partial U^*(p,r,z)}{\partial r}\right|_{r=0}=0; \tag{7}$$

$$\lambda\left.\frac{\partial U^*(p,r,z)}{\partial z}\right|_{z=0}=\begin{cases}-\varepsilon(1-e^{-t_2\cdot p})\cdot q^*(p,r) \text{ for } 0\le r\le R,\\ 0 \text{ for } r>R;\end{cases} \tag{8}$$

where

$$U^*(p,r,z)=\int_0^\infty e^{-p\cdot t}\cdot U(t,r,z)dt,\ q^*(p,r)=\int_0^\infty e^{-p\cdot t}\cdot q(t,r)dt,\ p>0.$$

We apply the integral Hankel transform with respect to the coordinate r to (6) - (8) and, using properties of the integral Hankel transform [8], we obtain a second-order differential equation in complete derivatives:

$$\frac{d^2\tilde{U}^*(p,\xi,z)}{dz^2}-\xi^2\cdot\tilde{U}^*(p,\xi,z)-\frac{p}{a}\cdot\tilde{U}^*(p,\xi,z)=0 \tag{9}$$

with boundary conditions

$$\tilde{U}^*(p,\xi,z)=0$$

for $z\to\infty$ and

$$\lambda \left. \frac{d\widetilde{U}^*(p,\xi,z)}{dz} \right|_{z=0} = -\varepsilon(1-e^{-t_2 \cdot p}) \cdot \widetilde{q}^*(p,\xi) \ , \qquad (10)$$

where $\widetilde{U}^*(p,\xi,z) = \int_0^\infty r \cdot U^*(p,r,z) \cdot J_0(\xi \cdot r) dr$, $\widetilde{q}^*(p,\xi) = \int_0^\infty r \cdot q^*(p,r) \cdot J_0(\xi \cdot r) dr$, $\xi \geq 0$ is Hankel integration parameter, J_0 is Bessel function of the first kind of zero order.

Solving equation (9) with conditions (10), and also, taking into account that the information is taken only from the surface of the body under investigation, for $z=0$ the dependence of the surface-time integral characteristic of the sample temperature $\widetilde{U}^*(p,\xi,0)$ with the surface-time integral characteristic of the heat flux $\widetilde{q}^*(p,\xi)$ can be found.

Solving equation (9) with boundary conditions (10), and applying method discussed in [9] an equation for calculating thermal diffusivity coefficient a, thermal conductivity λ and volumetric heat capacity cρ can be obtained:

$$a = \frac{p \cdot R^2}{g(p)} \qquad (11)$$

$$\lambda = \frac{2\varepsilon \cdot R \cdot q^*(p)(1-e^{-t_2 \cdot p})}{S^*(p)} V(g(p)) \qquad (12)$$

$$c\rho = \frac{\lambda}{a} \qquad (13)$$

where $g(p)$ and $V(g(p))$ are functions computed according to the method described in [9].

2.2. Development of automated control system

To implement the described method, we developed automated control system consisted of the functional blocks presented on the Figure 2. The laser had a maximum output power of 0.5 W and a wavelength of 405 nm. Focusing of the laser made it possible to obtain a light spot with a diameter of several tenths of a millimetre to 10 mm at a distance of 10 mm to 200 mm to the object surface. The laser power control unit allowed the output power controlled by pulse width modulation.

To obtain the primary thermal information from the body surface, a thermal imager manufactured by FLIR Model A35 was used. Then we designed thermal imager which generated a video stream at a rate of up to 60 frames per second. The software, which was developed in the LabVIEW 2016 graphical programming environment, received the image from the thermal imager, time-by-frame processes it using the NI Vision technology, and determined the integral temperature $S^*(p)$ on the circular object surface with radius R. The TPCs measured with the presented system were used in the automated device 8 (Figure 2) of an active potato quality control to determine the heat treatment regimes in automatic sorting complexes.

An algorithm of the experimental data processing is shown on the block diagram (Figure 3) and is described below.

IOP Conf. Series: Journal of Physics: Conf. Series **1084** (2018) 012002 doi:10.1088/1742-6596/1084/1/012002

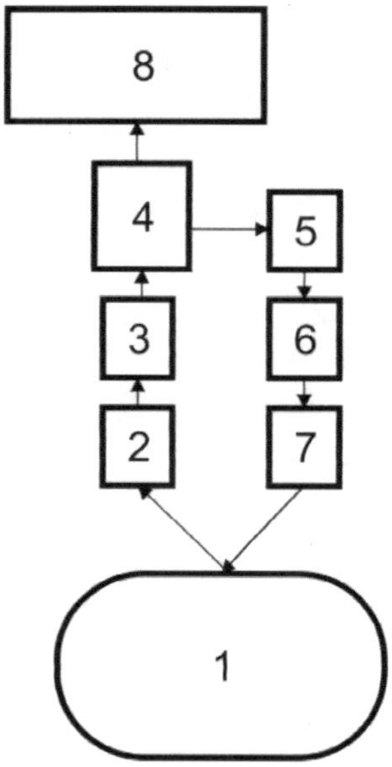

Figure 2. Measuring unit: 1 – object of control, 2 – thermal imager, 3 – adapter, 4 – personal computer, 5 – control unit, 6 – power supply, 7 – laser, 8 – automated device.

At first the image grabbing parameters must be set. These parameters include the camera model, resolution, temperature sensitivity, shooting speed. Then the correct port of the input-output board must be set in order to control the laser heating device. When the parameters are set a test sample should be heated by laser radiation. Region of interest (ROI) should be set by moving the cursor. Then from this ROI and the image a mask is created in order to take into account only temperature points in ROI. From this mask an array of temperature points is obtained. Then an integral surface characteristics in the ROI can be calculated for each period of time. These points are displayed at the thermogram graph. Image from the camera also is displayed. A test sample is heated until the temperature has reached the target value. Then the laser is switched off. After that the calculation step begins. A user should set the Laplace transform parameter p and the coefficient k. All the parameters and TPC are calculated. At the last step the calculated TPC are displayed.

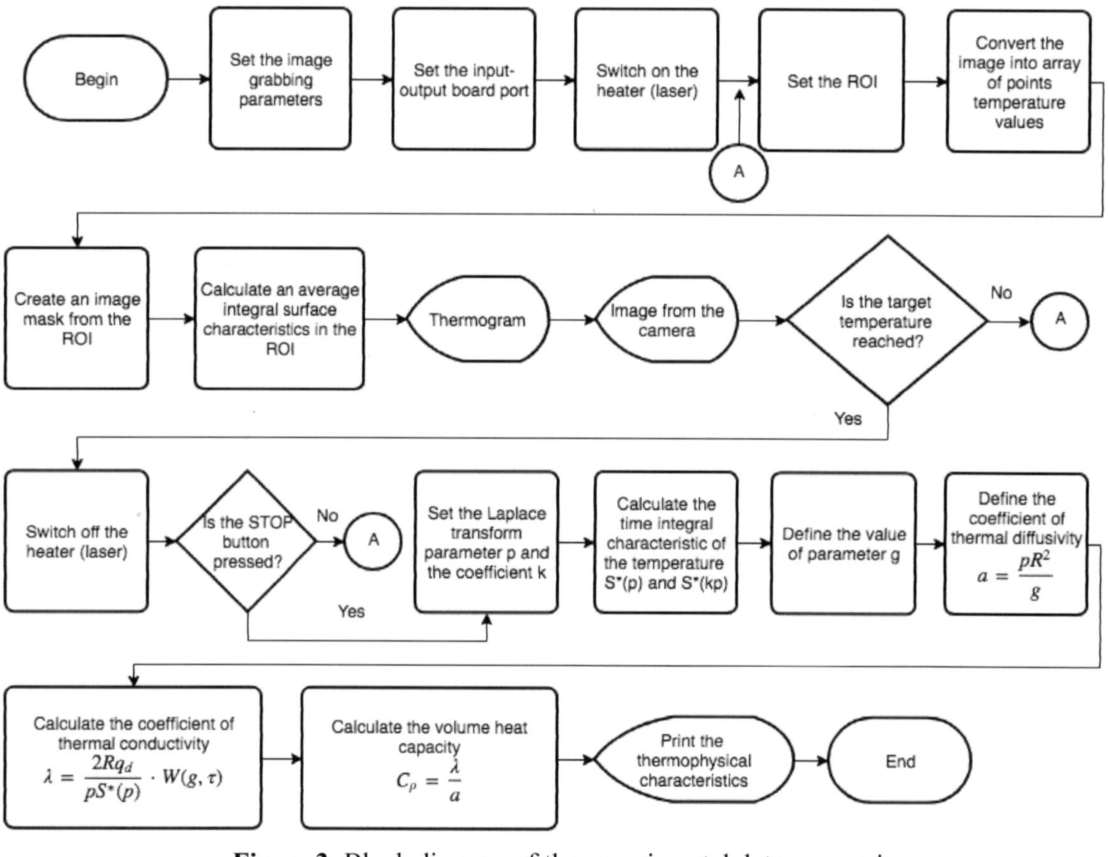

Figure 3. Block diagram of the experimental data processing.

2.3. Experimental approval of the developed approach

Using the presented measuring device, we carried out experiments on the tissue of an early-adapted sort of potato "Udacha" selected from the Tambov region, Russian Federation in 2017 year immediately after harvest. Both samples had known TPCs. The heat was produced by a laser beam at a maximum power of 0.5 W at a distance of 15 cm, focused on the surface area of the object with a radius of R=4 mm. We analyzed two samples of potatoes – healthy potato and rotten plant – potato after 10-days rot under the influence of high temperature and high humidity.

The thermal conductivity measurement error using the pilot plant was determined by comparison with the results of the measurements obtained with the certified ITP-MG4-100 instrument ("SKB Stroypribor" LLC, Russian Federation). The experimental samples consisted of four parallelepipeds cut from the potato tubers with a size of 50x50x15 mm. Four samples were combined into one with a size of 100x100x15 mm. Taking into account the basic relative error of 5% of the ITP-MG4-100 instrument, the limit of the thermal conductivity measurement error using the pilot plant was no more than 12%. The error source appears to be the part of the radiation reflection from the object surface. The influence of the radiation reflection can be reduced by refining the coefficient k in the formula (5).

Experiments with potato tissue have shown that the effective thermal conductivity of a healthy tissue differs significantly from a defective one, in particular from a tissue affected by dry rot (Table 1).

Table 1. Thermal conductivity of healthy and rotten potato tissues.

The test sample	λ, $W \cdot m^{-1} \cdot K^{-1}$
Healthy potato tissue	0.50 ± 0.06
Rotten potato tissue	0.40 ± 0.06

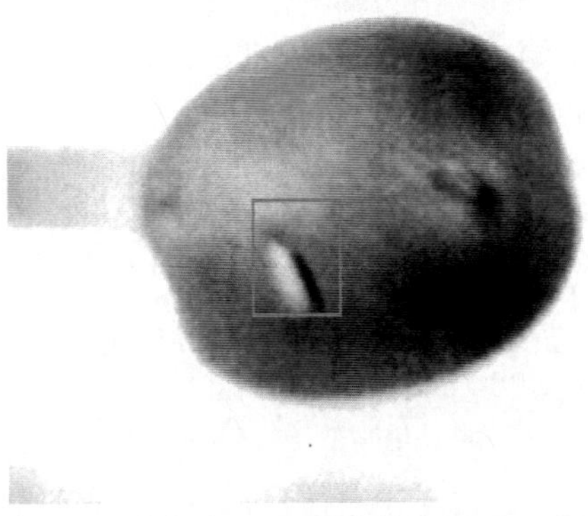

Figure 4. Potato with defect (the defective zone is highlighted with red rectangle).

The results of thermal conductivity measurement are indirectly confirmed by thermal imaging of potato tubers, which tissue has a dry rot type defect. The images clearly show the defective zones in the form of pixels sets with a different brightness (Figure 4). This fact allows using of NI Vision library standard tools for image processing and defect search.

Conclusion

The fact that plant thermal conductivity strongly depends on structural disturbances confirms the possibility of using thermal non-destructive control of plant tissues by selecting power and the duration of thermal exposure. In order to obtain a confidently detectable temperature contrast between the defective and defect-free zones of monitored samples, it is necessary to organize a thermal action of a certain power and duration as shown in the paper. At the same time, in the process of non-stationary heat exchange, a temperature difference arises on the surface of an object with defects, caused by the heterogeneity of its structure, and as a result, by unequal TPC. This temperature difference is the most informative parameter, according to which the presence of not only surface but also subsurface defects is judged. In order to determine thermal conditions, it is necessary to simulate non-stationary process of heat transfer, in a mode when thermal radiation with known characteristics acts on the monitoring object. The developed method for non-destructive testing of potato TPCs allows the use of thermal control to detect rotten tubers.

For a reliable detection of a rotten potato, it is necessary to obtain an image of its entire surface, which can be achieved by the use of multiple cameras, rotation of a sample at equal rotation speed and shooting frequency. In addition, the presence of thermal interference from the contact of the monitored object with the conveyor belt or rollers that rotate the object should be taken into account. The use of thermal control in conjunction with technical vision makes it possible to detect not only surface, but also sub-surface defects that cannot be detected visually.

Acknowledgments
The work was supported by the Ministry of Education and Science of the Russian Federation under the Agreement 14.577.21.0214 (Applied research and experimental development RFMEFI57716X0214).

References

[1] Is'emin R L, Kuz'min S N, Mishchenko S V, Divin A G, Milovanov O Y and Klimov D V 2015 Determination of temperature dependence of biomass heat capacity in low-temperature pyrolysis (prepyrolysis) conditions *Chem. Pet. Eng.* **51(1)** 6

[2] Ginzburg A S and Gromov M A 1987 *Thermophysical Characteristics of Potatoes, Vegetables and Fruits* (Moscow: Food Industry)

[3] Mishhenko S V, Divin A G, Gromov Yu Yu and Balabanov P V 2016 *Proc. Int. Conf. of Virtual Simulation, Prototyping and Industrial Design* (Tambov: Tambov State Technical University) pp 20–27

[4] Liaoa G, Wang X, Jin J and Li J 2015 Potato size and shape detection using machine vision *MATEC Web of Conf.* **31** 15003

[5] Pu Y Y, Feng Y Z and Sun D W 2015 Recent progress of hyperspectral imaging on quality and safety inspection of fruits and vegetables: a review *Compr. Rev. Food Sci. F. Safety* **14** 176

[6] Baranowski P, Mazurek W, Wozniak J and Majewska U 2012 Detection of early bruises in apples using hyperspectral data and thermal imaging *J. Food Eng.* **110(3)** 345

[7] Mishhenko S V, Churikov A A and Podol'skij V E 1989 Materials moisture non-destructive control *Equipment in Agricult.* **4** 35

[8] Watson G N 1966 *A Treatise on the Theory of Bessel Functions (2nd. ed.*) (Cambridge: Cambridge University Press)

[9] Gromov Y Y, Churikov A A, Divin A G, Ishchuk I N and Barkalov S A 2016 Hardware and software system for thermal control *J. Eng. Sci. Technol. Rev.* **9(6)** 135

IOP Conf. Series: Journal of Physics: Conf. Series **1084** (2018) 012003 doi:10.1088/1742-6596/1084/1/012003

Mathematical modeling of alkyde resin paint modified by carbon nanotubes

Artem Rukhov[1], E Tugolukov[2], S Juboori[1,3], K Brankin[1] and Anton Rukhov[4]

[1]Department "Chemistry and Chemical Technologies", Tambov State Technical University, 106 Sovetskaya street, Tambov 392000 Russia

[2]Department "Technology and Technology of Production of Nanoproducts", Tambov State Technical University, 106 Sovetskaya street, Tambov 392000 Russia

[3]Chemical Engineering Department, University of Technology, Mohamid PO Baghdad, 35023 Irak

[4]Department "Operation of Motor Transport and Car Service", Tambov State Technical University, 106 Sovetskaya street, Tambov 392000 Russia

[1]E-mail: rukhov.av@gmail.com

Abstract. The purpose of the paper is to develop a hypothesis of the alkyd resin modification process. Using the method of molecular mechanics, a mathematical simulation of the alkyd resin-carbon nanotubes system was carried out under equilibrium conditions. It is shown that the surface of carbon nanotubes is able to structure polymer chains with the formation of additional hydrogen bonds, that can explain the increase in hardness of the film by 10-60%, obtained experimentally. Practical recommendations on the development of industrial technology for the production of modified alkyd paints have been discussed.

1. Introduction

Primers are widely used to ensure adhesion and to reduce paint consumption for subsequent staining with enamels of wooden and metal surfaces applied in the construction, repair and decoration works [1, 2]. Among the requirements, priming compound should effectively treat the surface, totally close its pores, quickly dry, create highly stable film and should not shrink when processing and operating the surfaces. However, currently produced primers do not always provide desired depth of it' penetration into the treated surface, then, good adhesion to the enamel or varnish layer, high elasticity, and impact strength [1, 3].

Due to high mechanical strength, heat resistance, large specific surface, and hydrophobicity, carbon nanotubes (CNTs) may be considered as promising modifier of primer as shown before [4, 5]. The purpose of the modification is to improve the quality characteristics of the primer, for example, regulated by the documents GOST 25129-82 Primer GF-021 (Russia). For example, it is expected that the mechanical strength of CNTs embedded in the primer base will make it less sensitive to incision. Previously we have shown [6] that alkyd-based primer containing 0.01% wt. of CNTs has film hardness 10-60% higher than the primer without CNTs. However, there are still no data about mechanisms of the processes taking place on the molecular level when the priming compositions are modified with carbon nanotubes, that limit primer introduction into the practice. Meanwhile, in most cases it is not necessarily to carry out experiments to obtain data on matter' structure and composition, since it can be obtained on the basis of mathematical modeling. The purpose of our work is to propose a hypothesis about modification of alkyd resins, that would explain an increase of film hardness after modification of polymer.

2. Calculations

Despite many benefits of quantum-chemical methods of modelling [7], they need high power requirements to computer device and poorly used for considering molecular systems with more than 50 atoms. Contrary, reliable data about the structure of CNTs distributed in alkyd resins demand of examining system of dimension no less than $10^3$of atoms, that cannot be studied by quantum-chemical methods.

In our work we modelled the molecular system based on GF-021 primer (alkyd resin and solvent) modified with CNTs "Taunit-M" (NanoTechCenter Ltd., Tambov, Russia).Carbon nanotubes in alkyd resin was introduced in the following way. At room temperature, a suspension of nanotubes in a solvent with a concentration of about 0.1% of the mass was obtained. For high-quality dispersion, ultrasonic treatment was used, with a specific power of 18 W/g for 40 minutes. In the process of ultrasonic treatment the suspension was cooled with running water. The resulting suspension in different ratios described in [6] was introduced into the alkyd varnish and intensively moved in the homogenizer for 1.5 hours at a shaft rotation speed of 10,000 min^{-1}. During the mixing process, the modified varnish was cooled with running water. The systems was examined in equilibrium state by the help of the method of molecular mechanics (MM+), which generally treats atoms as Newtonian particles interacting with each other through potential fields given on the basis of the empirical laws of physics. Here, the potential energy of interaction depends on bond length, bond and torsion angles, and non-covalent interactions (including van der Waals forces, electrostatic interactions and hydrogen bonds). It should be noted that the molecular mechanics method is applicable if the De Broglie wavelength of an atom is much smaller than the interatomic distance, particularly, for particles with a weight more than 10 atomic units and temperature higher 10 K [8].

The force fields of the potential energy of a molecular system are generally given as the sum of pair atomic-atom interactions, which depend only on the distance between atoms [9, 10]. For calculating system potential energy we used MM+ method developed for organic molecules, that took into account the potential fields formed by all atoms of the calculated system and allowed flexible modifying the calculation parameters depending on the specific problem, which makes it, on the one hand, to be universal, and on the other it sharply increases the necessary resources in comparison with other methods of molecular mechanics.

In the MM + method the molecule potential energy is considered as follows:

$$U = U_b + U_v + U_\varphi + U_f + U_q + U_{LJ} + U_{hb} \tag{1}$$

where U_b – chemical bonds; U_v – valence angles; U_φ – torsion angles; U_f – flat groups and pseudorotation angles; U_{LJ} – Van der Waals contacts; U_q – electrostatic interactions; and U_{hb} – hydrogen bonds.

The energy of valence interactions and vibration energy of valence angles were described by parabolic potentials. The potential energy for torsion angles, plane groups and pseudorotation angles was given by the Fourier series [11, 12]. The interaction between atoms not connected by a valence bond (van der Waals interaction and the formation of hydrogen bonds) was described using the Lennard-Jones potential [13]. The potential energy of interaction of charged atoms was characterized by an electrostatic potential.

The formulation of the problem of calculating the geometry of the molecular system of alkyd resin-CNTs, which was in equilibrium at a local minimum of the value of the potential energy, had the following form:

$$\{\bar{r}_{n,m}\};\ n=1..\ N_a;\ m=1..3; \tag{2}$$

$$l_t = f_l(\bar{r}_{n,m});\ \alpha_h = f_\alpha(\bar{r}_{n,m});\ \varphi_g = f_\varphi(\bar{r}_{n,m});\ t=1..T;\ h=1..H;\ g=1..G. \tag{3}$$

where $\bar{r}$ – the matrix of the coordinates of N_a atoms in 3D space; l_t – length vector T of chemical bonds; α_h – vector of values of angles H of chemical bonds; φ_g – vector of values of torsion angles G of chemical bonds.

$$U = U_b(\vec{l}) + U_v(\vec{\alpha}) + U_\varphi(\vec{\varphi}) + U_f(\vec{\varphi}) + \\ + U_q(\bar{r}) + U_{LJ}(\bar{r}) + U_{hb}(\bar{r}) \rightarrow \min \quad (4)$$

$$U_b(\bar{l}) = \frac{1}{2}\sum_{t=1}^{T} K_{b\,t}(l_t - l_{0\,t})^2 , \quad (5)$$

where $K_{b\,t}$ – effective stiffness of the valence bond; $l_{0\,t}$ – equilibrium chemical bond length.

$$U_v(\vec{\alpha}) = \frac{1}{2}\sum_{h=1}^{H} K_{v\,h}(\alpha_h - \alpha_{0\,h})^2 , \quad (6)$$

where $K_{v\,h}$ – effective elasticity of the valence angle; $\alpha_{0\,h}$ – equilibrium value of the chemical bond angle.

$$U_\varphi(\vec{\varphi}) = \frac{1}{2}\sum_{g=1}^{G}\sum_{s=1}^{S} K_{\varphi\,g,s}\left[V_{g,s}\cos(s\,\varphi_g - \delta) + 1\right], \quad (7)$$

$$U_f(\vec{\varphi}) = \frac{1}{2}\sum_{g=1}^{G}\sum_{s=1}^{S} K_{f\,g,s}\left[V_{g,s}\cos(s_{g,s}\,\varphi_g - \delta) + 1\right], \quad (8)$$

where $K_{\varphi\,g,s}$, $K_{f\,g,s}$ – constants determining the heights of the potential barriers of dihedral angles; $V_{g,s}$ – contribution of the harmonic to the potential of the torsion angle ($-1 < V_{g,s} < 1$); $s_{g,s}$ – torsional barrier multiplicity; δ – phase shift; s – harmonic number; S – number of harmonics.

$$U_q(\bar{r}) = \sum_{i=1}^{N_a - 1}\sum_{j=i+1}^{N_a}\left[\frac{q_i\,q_j}{\varepsilon\,W_{i,j}}\right], \quad (9)$$

where q_i, q_j – partial charges on atoms; ε –medium dielectric permittivity; $W_{i,j}$ – distance between i and j atom, $W_{i,j} = \left|\vec{r}_i - \vec{r}_j\right|$.

$$U_{LJ}(\bar{r}) = \sum_{i=1}^{N_a - 1}\sum_{j=i+1}^{N_a}\left[\frac{A_{i,j}}{W_{i,j}^{\;12}} - \frac{B_{i,j}}{W_{i,j}^{\;6}}\right], \quad (10)$$

$$U_{hb}(\bar{r}) = \sum_{i=1}^{N_a - 1}\sum_{j=i+1}^{N_a}\left[\frac{A'_{i,j}}{W_{i,j}^{\;12}} - \frac{B'_{i,j}}{W_{i,j}^{\;10}}\right], \quad (11)$$

where $A_{i,j}$, $B_{i,j}$, $A'_{i,j}$, $B'_{i,j}$ – constants determining the depth of potential wells and the location of their minimum depend on the types i and j of the atoms involved in the interaction.

Initial conditions: $$\{\bar{r}_{n,m}\}_0 = R_{N_a,3} \quad (12)$$

Limitations: $$\{r_{n,m}\} \in \Omega_{N_a,3};\ \bar{l}_{t\min} \le \bar{l}_t \le \bar{l}_{t\max}; \\ \bar{\alpha}_{h\min} \le \bar{\alpha}_h \le \bar{\alpha}_{h\max};\ \bar{\varphi}_{g\min} \le \bar{\varphi}_g \le \bar{\varphi}_{g\max}. \quad (13)$$

It was necessary to find variable parameters as $\bar{r}$ – coordinates of atoms in three-dimensional space (2), under which the criterion of optimality (the potential energy of the system) (4) tended to a minimum under conditions (4) - (13).

The solution of problem (3) - (12) with allowance for the constraint (13) was obtained using the method of conjugate gradients by the Polac-Ribiera (Pshenichny) algorithm [14] implemented in the Gromacs software product.

3. Results and discussion

According to the analysis of calculation results, the surface of carbon nanotube placed in an alkyd resin under noted conditions has a structuring effect (Figure 1). The polymer changes its configuration and forms new intramolecular hydrogen bond in its chain (black arrow on the Figure 1). The formation of a hydrogen bond is determined by the condition that the distance from electronegative atom to hydrogen donor is less than 3.2 Å, and the angle of the covalent bond between donor and acceptor is less than 120°. There are many polymers which properties are determined by the presence of hydrogen bonds, as cellulose, nitrocellulose, and nylon [15]. In alkyd resin this bond has an energy of about 20-25 kJ/mol and most likely performs an auxiliary function. Thus, it is possible to explain hardness growthof modified primer film by forming an additional hydrogen bond in the polymer. When the film is bent or stretched, this bond is easily torn by allowing polymer to rearrange. After removing the load, in the presence of favourable conditions, this hydrogen bond is formed again. This explains why the hardness of film increases although its fragility does not augment.

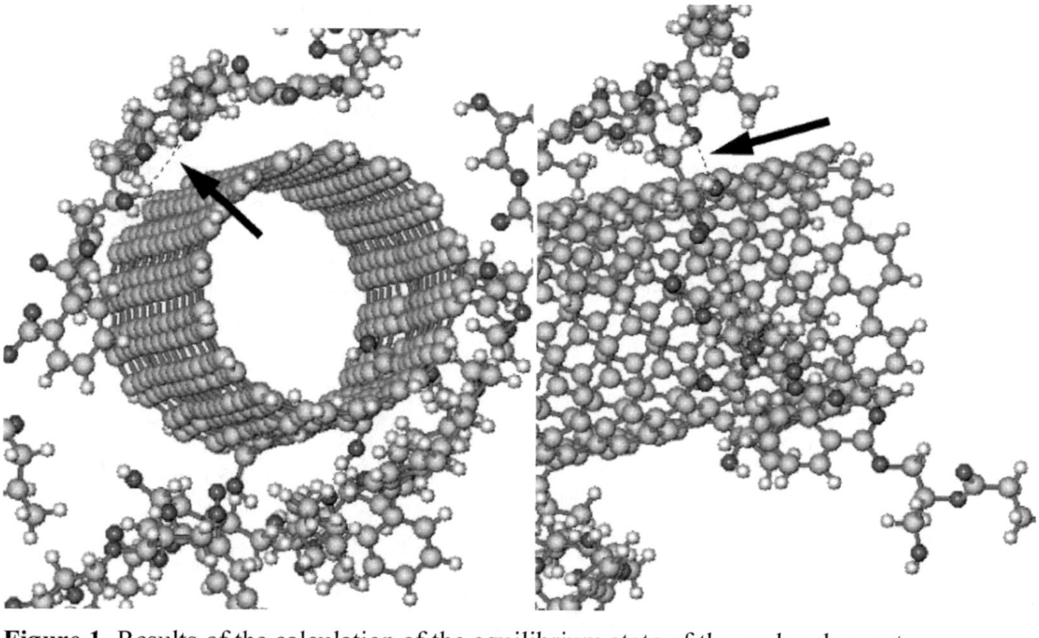

Figure 1. Results of the calculation of the equilibrium state of the molecular system of alkyd resin-CNTs. Arrow shows a new hydrogen bond.

There are several possible options for the practical implementation of this method of modifying alkyd resins by modernizing the traditional technology for the production of alkyd primers, typically in periodic synthesis of alkyd glythaldehyde primers shown in Figure 2.

In order to improve GF-021 primercharacteristics, it is proposed to introduce an additional stage into the traditional technological scheme by including the step of CNTs solvent modification. The hardware design of the new stage includes container for CNTs, dispenser, container for mixing CNTs in a solvent and ultrasonicator providing uniform distribution of nanotubes in a solvent.

The problem of agglomeration of CNTs in an organic solvent is still unsolved [16], however, as research shows [17] application of hydrophobic surfactants reduces nanotubes aggregates since they absorb on the phase interface. So, it is advisable to equip a new stage also by feeding the surfactant into the slurry prior to its ultrasonic treatment. Another variant to improve traditional technology for primer preparation may involve the introduction of CNTs during preparation of dry mixture of fillers, chalk and microtalc, followed by its processing in inkjet machine. In this case, in the technological scheme it is necessary to install only container for CNTs and dispenser. It was also considered a variant of technology modernization - introduction of CNTs into semi-finished alkyd varnish.

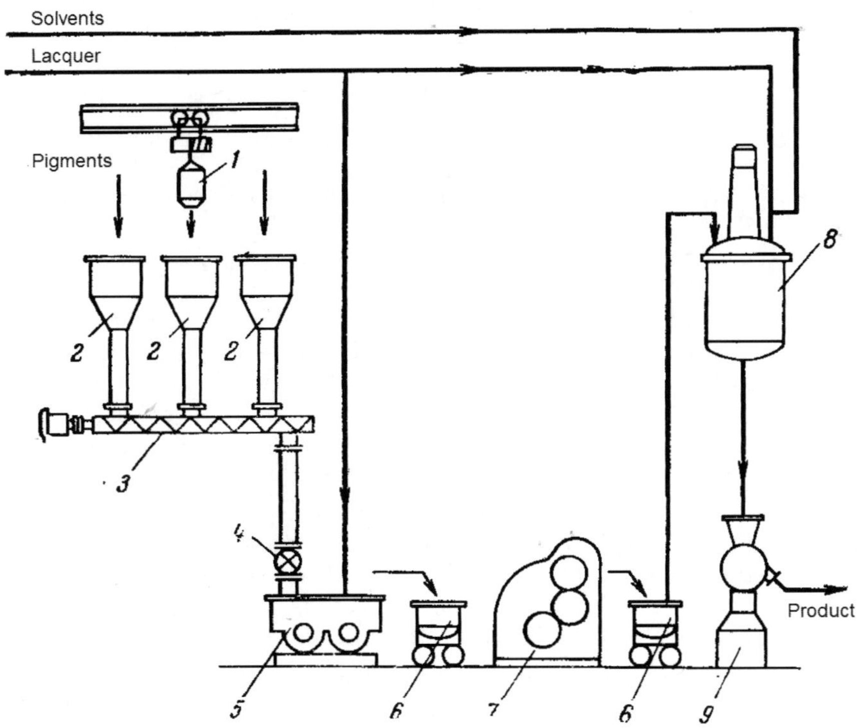

Figure 2. Technological scheme of production of alkyd glyptal primers in a periodic way: 1 – container for pigments; 2 – hopper-funnel; 3 – auger; 4 – feeder; 5 – kneading machine; 6 – container for mixing; 7 – roller mill; 8 – mixer; 9 – colloid mill.

Conclusions

Using the method of molecular mechanics (MM+), the mathematical modeling of the molecular system of alkyd resin-CNT in equilibrium conditions has been carried out. For the first time it is proved that in certain configurations the surface of carbon nanotubes is able to structure alkyd polymer chains with the formation of intramolecular hydrogen bonds.

The calculated data on the formation of hydrogen bonds with energy of 20-25 kJ/mol explains the increase in the film hardness by 10-60%, obtained experimentally earlier [6]. The formation of additional bond in the polymer allow us to propose a hypothesis about the processes of alkyd paint modification by carbon nanotubes. When the film is compressed or stretched, this bond is easily broken allowing the polymer to rearrange. After removing the load, in the presence of favorable conditions, this hydrogen bond is formed again. As the hardness of the film increases, its brittleness remains constant. The obtained results of modeling and the formulated hypothesis allow us to follow the conclusion about the high prospects of alkyd composition modification by carbon nanotubes. In this regard, we propose to introduce the process of alkyd modification by CNTs in industrial scale, that opens up prospects for the creation of new durable paints. The formulation of the hypothesis about the modification processes requires its comprehensive experimental confirmation. Future works of the

IOP Conf. Series: Journal of Physics: Conf. Series **1084** (2018) 012003 doi:10.1088/1742-6596/1084/1/012003

authors will be focus on the experimental study of the alkyd paints modification with carbon nanotubes.

Acknowledgments
This research is supported by the grant of Tambov region (Russia) for young scientists in 2017, agreement No. 1-17/MU, project No. 14 MU-17.

References
[1] Elrebii M, Kamoun A and Bouf S 2015 Waterborne hybrid alkyd–acrylic dispersion: Optimization of the composition using mixture experimental designs *Prog. Org. Coat.* **87** 222
[2] Berge C B and Henley A 2009 Paint, varnish, stain and wax *The Ecology of Building Materials* **18** 381
[3] Abdel N A and Madkour M Potential use of smart coatings for corrosion protection of metals and alloys: a review *J. Mol. Liq.* **253** 11
[4] Dyachkova T, Anosova I, Galunin E, Orlova N and Tkachev A 2017 Preparation of TiO_2/carbon nanotubes composites and a study of their adsorption on organic dyes *Nano Hybrids and Composites* **13** 135
[5] Burakova E, Galunin E, Rukhov A, Memetov N and Tkachev A 2016 Effect of ultrasound on a mixed oxide-based catalyst in the synthesis of carbon nanostructured materials *Res. Chem. Intermed.* **42** 7045
[6] Filatova E and Rukhov A 2011 Study of soil compositions modified by carbon nanostructured material Taunit-M *Prov. Int. Conf. "Perspective Innovations in Science, Education, Production and Transport 2011",* 21-30 June, 2011, Odessa, Ukraine
[7] Clark T 1985*A Handbook of Computational Chemistry: A Practical Guide to Chemical Structure and Energy Calculations* (Wiley, Moscow) p 383
[8] Rapaport D C 2004 *The Art of Molecular Dynamics Simulation* (Cambridge University Press), p 565
[9] Mc Cammon J A 1984 Protein dynamics *Rep. Progr. Phys.* **47**1 46
[10] Kalinichev A G and Kirkpatrick R J 2002 Molecular dynamics modeling of chloride binding to the surface of calcium hydroxide, hydrated calcium aluminate, and calcium silicate phases.*Chem. Matter.* **14** 3539
[11] Tormena C F 2016 Conformational analysis of small molecules: NMR and quantum mechanics calculations *Progress in Nucl. Magn. Resonance Spectroscopy* **96** 73
[12] Peng-Yuan C, Lin Z, Shun-Guan Z and Guang-Bin C 2017 Role of intermolecular interaction in crystal packing: A competition between halogen bond and electrostatic interaction *J. Mol. Struct.* **1131** 250
[13] Kaplan I G 1982 *Introduction to the Theory of Intermolecular Interactions* (Nauka, Moscow) p 312
[14] Nguyen A, Reiter S and Rigo Ph 2014 A review on simulation-based optimization methods applied to building performance analysis*Appl.Energ.* **113** 1043
[15] Raj P H, Bajgai M P, ChuanY, Nirmala R, Nam K, Baek W and Kimd H Y 2010 Effect of successive electrospinning and the strength of hydrogen bond on the morphology of electrospun nylon-6 nanofibers *Colloid Surface A* **370** 87
[16] Piao-Hao S, Kim-Hwan M, Choi-Jin H, Nirmala R, Lee H and Park J *J. Ind. Eng. Chem.* 2017 Dispersion state and rheological characteristics of carbon nanotube suspensions **52** 369
[17] Nomède-Martyr N, Disa E, Guérin K, Bonnet P and Dubois M 2018 Effect of fluorination on the stability of carbon nanofibres in organic solvents *Cr. Chim.* **21** 791

Design and technological solutions for technical systems based on the example of a sowing complex for grain crops

K Nemtinov[1], A Eruslanov[1] and Y Nemtinova[*1, 2]

[1]Department of «Computer-integrated systems in mechanical engineering», Tambov State Technical University, 106 Sovetskaya Street, Tambov 392000, Russian Federation
[2]Department of «Management, marketing and advertising», Tambov State University named after G.R. Derzhavin, 33, Internatsionalnaya Street, Tambov 392000, Russian Federation

[1*]E-mail: jnemtinova@hotmail.com

Abstract. In this paper, we have developed a formalized problem statement for automated design of a technical system (TS) on the example of a sowing complex (SC) and improved the methodology of decision-making information support for its design by creating a virtual model of the TS. An information-logical model, which unifies sets of data on the external environment and data processing technologies, is used for selecting the structural scheme of the TS. A procedural decision-making model allows optimal selection of standard units and mechanisms of the TS taking into account uncertainty and discrepancies in the data. The virtual model of the SC is presented in the form of a hierarchical frame structure, including parametric three-dimensional geometric models and drawings of individual units, devices, mechanisms and details, as well as attributive data. A new design of a SC for grain crops uses both standard units and non-standard devices, which provides reduction of the sowing time and energy consumption together with the possibility of using the SC as a cultivator and a disk harrow.

1. Introduction

Development of decision-making control systems for design of technical systems (TS) is determined by intensive development of methodological foundations of design, mathematical models and algorithms, software and hardware. This problem is essential for making decisions concerning the structure of TS and its specific design parameters, in particular for the agro-industrial complex (AIC) of the Russian Federation [1-5].

The analysis of literature sources on the technology of automated TS design (using the example of sowing complexes (SC)) showed that much attention is paid to structural design of individual TSs or their nodes, and to a lesser degree to optimal selection of TS's structural scheme from the standpoint of different criteria (economic criterion, reliability, manufacturability, etc.) tied to specific initial data defined by technical specification and consumers: type of grain crops, soil type and moisture, soil tillage technology, drawbar category of the tractor, etc. [6-12]. As a result, it is not possible to consider all available variants of TS's structural scheme and to choose an optimal one from the standpoint of different criteria.

By investigating the technologies of computer-aided design of complex TS, it was concluded that the whole set of problems solved at all stages of TS design should be considered using the theory of

IOP Conf. Series: Journal of Physics: Conf. Series **1084** (2018) 012004 doi:10.1088/1742-6596/1084/1/012004

complex or large systems [1, 8, 13]. Such systems consist of many separate subsystems, each of which solves its own tasks, in particular: obtaining information about the object, transforming it into appropriate control law and transferring it to the object. The breakdown of a complex control system into separate subsystems is stipulated by large dimension of such systems and difficulties associated with gathering and processing of information about their state, and also selecting control actions. In most cases connection between individual subsystems is hierarchical.

Hierarchical systems are generally characterized by:

- location of subsystems with characteristic local properties in accordance with the levels of adopted hierarchy (in the overall structure of the system) corresponding to significance of decisions they make;
- implementation of connections from subsystems of lower levels to subsystems of higher levels by transferring generalized information;
- implementation of connections from subsystems of higher levels to subsystems of lower level through control actions.

In solving a number of problems of technical objects' design it is expedient to use expert systems that are used for hard-to-formalize and unformalized tasks, which are characterized by the following features [7-8]: incompleteness and inconsistency of initial data and knowledge about the research domain; large dimension of the problem.

In this regard, we have proposed an improvement of decision-making system for design and technological solutions for TSs the ultimate goal of which is to reduce the cost of design and to improve the quality of TS projects as well as to increase the efficiency of their operation on the example of a SC. To achieve this goal it is necessary to solve the following tasks:

- carry out a formalized statement of the decision-making problem for the system of automated TS design using the SC (sowing complex) example;
- improve the methodology of information support for making decisions on TS design;
- develop models for decision-making support in selecting of TS's structural scheme, as well as its typical operating bodies, devices and mechanisms;
- develop an original design of a SC for grain crops, including a seeder and a disk harrow, using special software;
- carry out experimental studies of the developed SC and evaluate its effectiveness.

The obtained results are based on the achievements of various scientific schools in the field of systems analysis, management, data processing, and technological design for the agro-industrial complex [6-12].

2. Purposes and objectives of research

The purpose of this work is to reduce the cost of development and to improve the quality of TS's projects, as well as to increase the efficiency of their operation on the example of a SC.

In order to meet these goals it is necessary to solve the following research tasks:

- perform the analysis of existing methods of decision-making information support for the structural design of TSs;
- develop a formalized problem statement of the automated TS's design on the example of a SC;
- improve the methodology of information support for making TS's design decisions;
- develop mathematical models of decision support for selection of a structural scheme, as well as its standard operating units, devices and mechanisms;
- develop a virtual model of TS with the reference to the SC;
- develop the structure of information support and software for the system of automated TS design on the SC example;
- develop the original design of a SC for grain crops, including a seeder and a disk harrow;
- carry out experimental studies of the developed SC and estimate its economic efficiency.

3. Problem statement of technical system design and decision support model

In general, the problem of automated TS design for a set of specific initial data (with the reference to the SC it includes the type of grain crops, soil type and moisture, soil tillage technology, drawbar category of the tractor and criteria set by the consumer) provides: development of its structural scheme, selection of typical elements, and design of original units, devices and mechanisms.

In a formalized form, the problem of TS design is to find an optimal value (minimum) of a function:

$$I_{opt} = \min_{w \in W} F_w \,, \tag{1}$$

under the constraints:

$$tx_i^{\min} \le tx_i \le tx_i^{\max}, \; i = 1,2,\ldots N_k, \; tx_i \in Tx \,, \tag{2}$$

and operators representing mathematical models of decision support for selecting:

- structural scheme of a TS:

$$\Psi : Pu \times Va \times Zb \times Rm \times Du \times Tu \to K \,, \tag{3}$$

- typical elements of a complex (units, devices and mechanisms):

$$\Phi : Tx \times R \times G \times k_{opt} \to Kt \,, \quad k_{opt} \in K, \tag{4}$$

- design solutions for TS as a whole:

$$\Omega : kt_{opt} \times k_{opt} \times P^3 \to Ku \,, \; kt_{opt} \in Kt, \; k_{opt} \in K \,, \tag{5}$$

- design values and operating conditions of individual elements of TS as a whole:

$$\Lambda : ku_{opt} \times k_{opt} \times kt_{opt} \to Tx, \; kt_{opt} \in Kt, \; k_{opt} \in K, \; ku_{opt} \in Ku \,, \tag{6}$$

where: F_w – generalized criterion for the w-th version of the TS; K, Kt, Ku – sets of variants for TS's structural schemes, typical units, devices and mechanisms, and design solutions, respectively; k_{opt}, kt_{opt}, ku_{opt}, tx_{opt} – optimal choice of a structural diagram; typical units, devices and mechanisms; TS's design; design values and operating parameters of individual elements, respectively; N_k - number of technological parameters; $tx_i^{\min}$, $tx_i^{\max}$ – the minimum and maximum values of the i-th technological parameter, $tx_i \in Tx$; Tx – set of parameters; P^3 – three-dimensional space for placement of units and mechanisms; Pu, Va, Zb, Rm, Lu, Tu – sets of main nodes types; $W = K \times Kt \times Ku \times Tx$ – set of TS variants; $w_{opt} = \{k_{opt}, kt_{opt}, ku_{opt}, tx_{opt}\}$, $kt_{opt} \in Kt$, $k_{opt} \in K$, $ku_{opt} \in Ku$, $tx_{opt} \in Tx$ – an optimal version of TS's project; Ψ, Φ, Ω, Λ – functional operators.

It is almost impossible to obtain the solution to this problem due to high dimensions of variables sets, complexity of mathematical models for design decision-making, complexity of formalizing individual problems, specific design features of individual units, devices and mechanisms of the complex. In this regard, the proposed decision-making method for TS design consists of a sequential solution of 4 sub-tasks of smaller dimension, interconnected by information flows, and possessing independent significance in the design process. In Figure 1 it is shown a conceptual model of the decision-making process for TS design on the example of a SC.

The problem A_s of automated formation of admissible TS structural schemes in accordance with the project specifications and selection of an optimal one based on a vector criterion is rarely solved. This is due to various difficulties, e.g. formalization of the problem. In order to solve such problems expert systems are most often used.

In order to solve this problem a structured database was created. Data structure for our subject area is presented in the form of its virtual model.

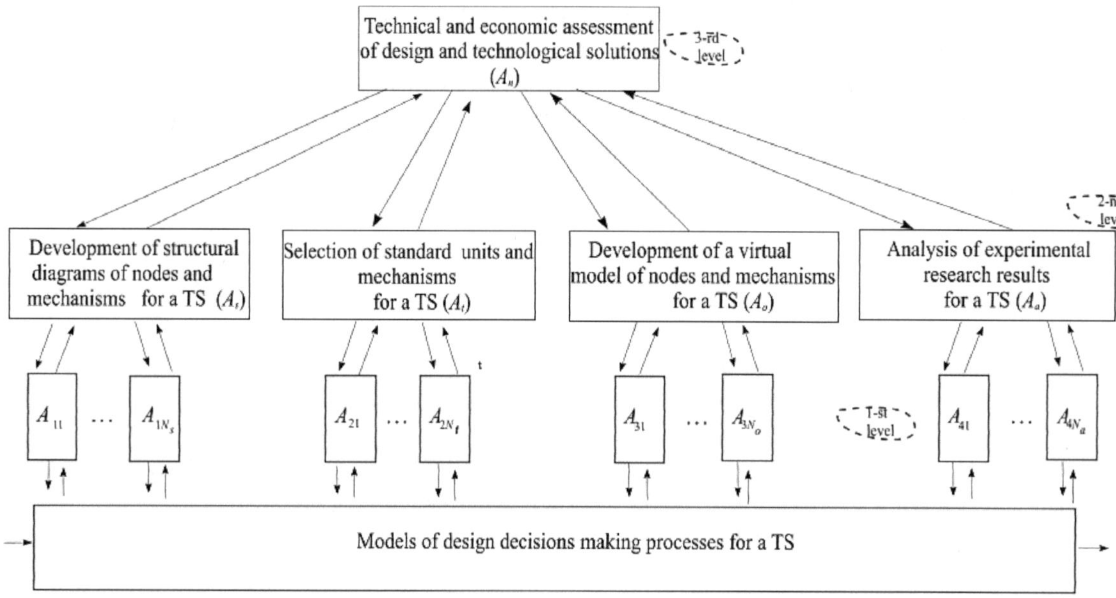

Figure 1. Conceptual model of decision-making process for TS design on an example of SC.

Data structure in the TS's virtual model - TS_{em}, describing the whole data set:

$$TS_{em} = \{Sc_{em}, P_{em}, Ms_{em}, Ma_{em}, Mg_{em}, D_{em}\},$$

where Sc_{em} - a frame describing the structural composition of a TS; P_{em} - a frame describing the properties of a TS as a whole; Ms_{em} - set of ways to determine the properties of a TS; Ma_{em} - set of attributive characteristics of a TS; Mg_{em} - set of parametric graphic models of a TS; D_{em} - a set of two-dimensional drawings of a TS.

In order to form the functional structure of TS we used production rules of the following type: if … (fulfillment of a condition), then … (implementation of a sequence). Below there is an example of such rules applied to the SC design problem:

- if (frame = "solid" and tanker type = "combined"), than dispenser type = "roller, quantity =1";
- if (frame = "solid" and tanker type = "separated for four dispensers"), than dispenser type = "roller, quantity = 4";
- if (frame = "solid" and tanker type = "combined" and dispenser type = "roller, quantity =1"), than transportation unit = "screw conveyer";
- if (frame = "solid" and tanker type = "combined" and dispenser type = "roller, quantity =1"), than transportation unit = "bucket elevator";

At present, the dimension of the set of SC structural schemes does not exceed a thousand, thus the method of full enumeration can be applied to find an optimal solution.

The next problem in making TS's design decisions on the example of a SC for grain crops is selection of typical elements (units, devices, mechanisms, etc.) from the assortment manufactured by the industry - A_t. In most cases, a significant part of them have a standardized size (hoe and chisel colters, bearings, disc knives, etc.). On the other hand, grain tankers, packing and transporting devices are manufactured according to individual orders.

Due to a wide variety of parts, mechanisms and units of agricultural machinery which have the same basic parameters, it becomes necessary for designers to select the one (unit or mechanism) with characteristics corresponding to a separate subset R^{zad} of specific customer requirements [13-16].

The authors propose a decision-making model for the selection of a typical part (unit, device or mechanism) of TS that ensures selection of the best one from the standpoint of the accepted criterion, that is characterized by the best technical characteristics and consumer properties obtained on the basis of quantitative (not always accurate) or qualitative information.

The decision support model for the selection of typical units, devices and mechanisms of the TS includes the process of forming a set of types of parts (units, devices or mechanisms) for the required used satisfying the values of technical characteristics. Further, based on the full enumeration method an optimal type is chosen.

4. Implementation of mathematical models of decisions support for design of a technical system

Implementation of the information support methodology for decision-making in design of technological complexes was carried out at LLC CB "ERUSLAN" (Russia) on the example of SC for grain crops. The initial data for design: soil tillage technology = "zero"; sowing material = "grain crops, wheat"; drawbar category of the tractor <= 1.4; soil type = "chernozem (humus content 5-9%, acidity ~ 7 pH)".

As a result of solving the A_s problem a technological scheme of the SC was chosen, including the following main operating units, devices and mechanisms: frame = "solid", tanker type = "combined", packing device = "pneumatic tires", sowing mechanism = "disk knives", dispenser type = "roller", transportation unit = "bucket elevator".

As result of solving the A_t problem for the SC the following industry-standard typical nodes were chosen: pneumatic tires brand = "disks from the UAZ452 vehicle, tires 235/65 / R16" quantity of 15 pcs.; disk knives = "BDT smooth disk with a diameter of 560 mm" quantity of 42 pcs.; dosing unit brand "SZP".

The results of virtual models development for non-standard units and mechanisms (solutions of the problem A_o): frames of the grain tanker and bucket elevator are shown in Figure 2.

Visualization of the virtual model and a photo of the prototype are shown in Figure 3 and Figure 4.

5. Analysis of experimental results for the developed technical system

Based on design features of the developed SC we have done the adjustment of the optimal values of its design parameters for a disk knife. In the course of experimental studies the angle of inclination of the disk varied from 10 degrees to 20 degrees. As a result of the experiment required data was obtained and correlation dependences between the value of specific traction resistance R (R_{Cc} - on the stubble background, R_{Cn} - on the fallow background) and the studied parameter were represented by the second degree polynomials.

As a result of experimental data approximation done with the least squares method we have obtained the regressions of specific traction resistance on the disk inclination angle l_r.

Optimal (minimal) values of these functions are achieved at the disc angle of 15 degrees, while the specific traction resistance of the SC on the stubble background does not exceed 3.44 kN/m, and on the fallow background does not exceed 3.19 kN/m. This angle of inclination and was adopted in the final version of the SC design.

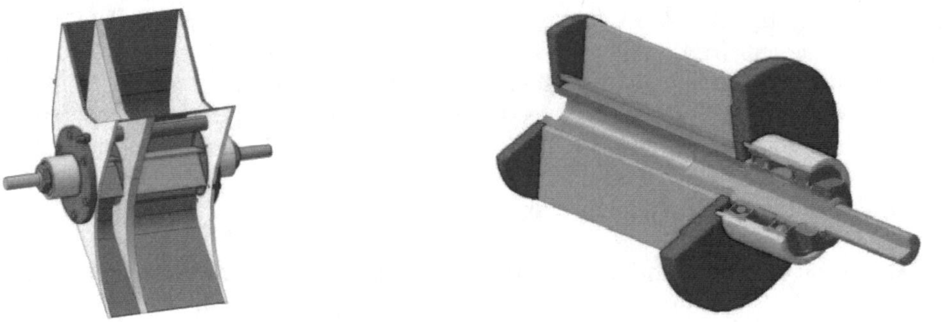

a) b)

Figure 2. Visualization of virtual models: a - roller dispenser; b - roller of the dispenser.

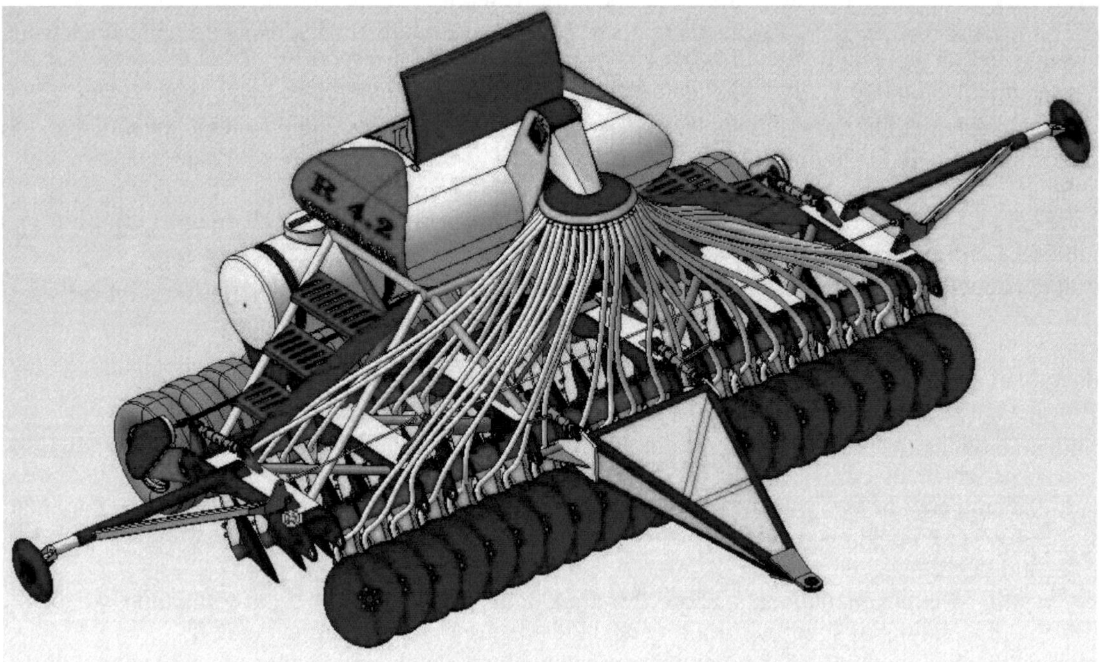

Figure 3. Visualization of the virtual model of the SC.

IOP Conf. Series: Journal of Physics: Conf. Series **1084** (2018) 012004 doi:10.1088/1742-6596/1084/1/012004

Figure 4. Photo of the SC prototype.

The optimal number of plants of grain crops per unit area of planting is largely related to the functioning of the bucket elevator and is provided by varying the following parameters: the volume of the buckets, the distance between the buckets and the speed of the bucket elevator.

The optimal values of these parameters were determined as a result of a computational experiment using the orthogonal central composition plan of the experiment. The bucket capacity (x_1) varied in the range from 3.0 dm^3 to 9.5 dm^3; the distance between the buckets (x_2) was in the range from 0.1 m to 0.25 m; and the speed of the elevator (x_3) - in the range from 0.047 m/s to 0.16 m/s.

As a response function, we used the function: $f(x_1, x_2, x_3) = | N_v - N_{\exp,i} |$, where: N_v – sowing rate per 1 m^2, $N_{\exp,i}$ - number of seeds during the i-th experiment, i=1, 2, ...,15. Variation levels of selected factors are given in Table 1. For three factors, correction of their squares a = 0.73, "star shoulder" α = 1.215. Experimental results are given in Table 2.

Optimal quantity of plants of grain crops per unit area of sowing is provided with the following parameters: bucket volume - 7,2 dm^3, distance between the buckets - 0.21 m; the speed of the bucket elevator - 0.12 m/s. As the degree of polynomial increases, their values remain unchanged.

Table 1. Variation levels of selected factors.

Factor	-α	-1	0	+1	+α
x_1	3.0	3.48	6.25	8.83	9.5
x_2	0.1	0.11	0.175	0.236	0.25
x_3	0.047	0.057	0.104	0.15	0.16

Table 2. Experimental results.

Level of factor x_1	Level of factor x_2	Level of factor x_3	$\lvert N_v - N_{\exp,i}\rvert$
-1	-1	-1	152
1	-1	-1	141
-1	1	-1	145
1	1	-1	96
-1	-1	1	76
1	-1	1	56
-1	1	1	24
1	1	1	15
-α	0	0	56
α	0	0	89
0	-α	0	66
0	α	0	97
0	0	-α	98
0	0	α	102
0	0	0	29

In assessing the performance of the SC, the following values were determined: specific traction resistance; the proportion of seeds in the soil layer; uniformity of distribution of wheat seeds in the area of sowing.

Experiments in laboratory conditions and field studies allowed compiling a list of advantages of the developed design of a sowing complex presented in Table 3.

Table 3. Results of structural analysis.

Indicator	Type of seeder	
	Base	Designed SC
Number of plants per 1 m, pcs	329	372
Number of productive plants, pcs	302	347
Plant height, cm	40.3	43.9
Length of a spike, cm	4.4	5.1
Number of grains in a spike, pcs	14.1	15.5
Grains weight of 1000 pcs	42.2	44.1
Grain weight from 100 plants, g	59.8	64.8
Biological yield, centner/ha	39.6	43.8

As an example, the results are given below: the depth of sowing (see Figure 5) and the distribution of seeds along the width of the sown band (see Figure 6).

6. Results and discussion

As a result of the presented research work the authors have solved an important scientific and practical problem of developing a decision management system for structural design of TS, implemented on the example of the original SC, which provides a cost reduction of up to ~ 15% and an increase in operating efficiency of ~ 17% in comparison with analogs.

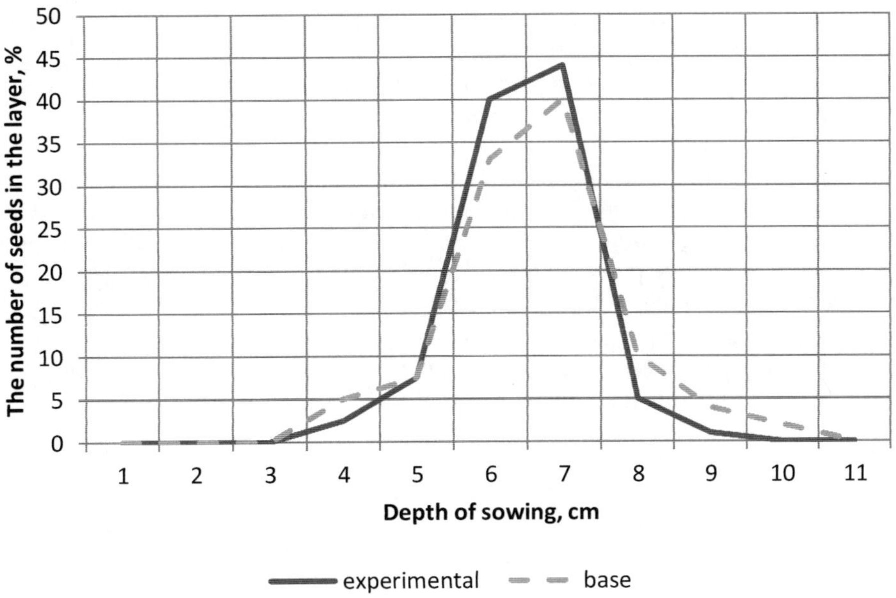

Figure 5. Results of sowing depth evaluation.

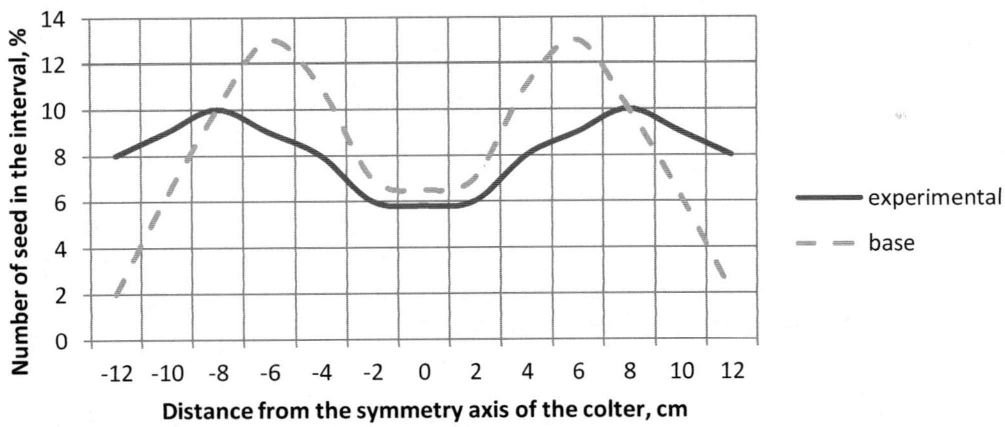

Figure 6. Results of evaluation of seeds distribution over the width of the sown band.

Other notable results include:
- reduction of sowing works duration as there is no gap between soil preparation and sowing;
- reduction of energy consumption by combining the operations of soil preparation, processing, sowing and packing of soil in one device (the use of diesel fuel for the whole cycle of sowing works is no more than 4 l/ha compared to the cost of performing these operations separately ~ 5-7 l/ha);
- optimal depth of sowing at the level of dew point, preserving the initial states of the capillaries, which ensure the natural supply of moisture from the ground;
- the possibility of using the SC as a cultivator and a disk harrow.

IOP Conf. Series: Journal of Physics: Conf. Series **1084** (2018) 012004 doi:10.1088/1742-6596/1084/1/012004

7. Conclusion

Based on the research results the following conclusions are made:

- a formalized statement of decision-making problem for computer-aided design of TSs on the example a SC was performed and the methodology of decision-making information support was improved taking into account TS's construction characterized by complexity of its structure, which includes standard and customized components and mechanisms; in addition a thorough analysis of computational and field experiments was conducted in order to clarify the values of individual structural elements of the TS;

- it was developed an information-logical model, which represents a combination of data sets containing information about external environment, technologies of its processing, and is crucial for making decisions about TS's structural scheme. A distinctive feature of the decision support model is its structure, which accumulates the set of relationships between all the main operating units and mechanisms of the TS with the help of production rules;

- a virtual model of the TS with the application to the SC is presented in the form of hierarchical frame structure, including parametric three-dimensional geometric models and drawings of individual operating units, devices, mechanisms and parts, as well as related attributive data.

References

[1] Nemtinov V A and Nemtinova Y V 2005 *Journal of Computer and Systems Sciences International* **44(3)** 389-398

[2] Mokrozub V G, Nemtinov V A and Mokrozub A V 2017 *Chemical and Petroleum Engineering* **53(5-6)** 326-331

[3] Lijing L 2015 *Transactions of the Chinese Society of Agricultural Engineering* **31(11)** 40-45

[4] Mokrozub V G and Nemtinov V A 2015 *Chemical and petroleum engineering* **51(7-8)** 31-35

[5] Lin J, Qian Wei, Li Baofa and Liu Yanfen 2015 *Transactions of the Chinese Society of Agricultural Engineering* **31(9)** 19-24

[6] Prasanna Kumar G V, Srivastava B and Nagesh D S 2009 *Computers and Electronics in Agriculture* **65(1)** 26-35

[7] Wang J M, Zhang L, Liu Y B, Mo X N and. Ren G Q 2010 *Computer Integrated Manufacturing Systems* **16(10)** 2017-2023

[8] Mokrozub V G, Manuilov K D, Gorshkov V V and Gorshkova T S 2016 *Chemical and Petroleum Engineering* **51(9-10)** 613- 617

[9] Ivannikov A, Kulagin V, Romanov A and Pozdneev B 2016 *Proceedings* of *IEEE EAST-WEST DESIGN & TEST SYMPOSIUM* (*EWDTS'2016*) (Yerevan, October, 14-17, 2016) 248-251

[10] Pappalardo C M and Guida D 2018 *Machines* **6(2)** 19

[11] Pappalardo C M and Guida D 2017 *Meccanica* **52(11-12)** 2503-2526

[12] Pappalardo C M and Guida D 2017 *Journal of Dynamic Systems, Measurement, and Control* **139(8)** 081010

[13] Malygin E N, Mokrozub V G and Nemtinov V A 2017 *MATEC Web of Conferences* **129** 01009

[14] Vamerali V, Bertocco M and Sartori L 2006 *Soil & Tillage Research* **89(2)** 196-209

[15] Yang L, He X T, Cui T, Zhang D X, Shi S, Zhang R and Mantao W 2015 *International Journal of Agricultural and Biological Engineering* **4** 1-9

[16] Zhao T, Zhao Y, Higashi T and Komatsuzaki M 2012 *Engineering in Agriculture, Environment and Food* **5(2)** 50-56

Virtual modelling of particles two-step feeding

T Al Sayyad[1], V Pershin[1], A Pasko[2] and T Pasko[2]

[1]Mechanical Department, Tambov State Technical University, 106 Sovetskaya Street, Tambov 392000, Russian Federation
[2]Nanomaterials Department, Tambov State Technical University, 106 Sovetskaya Street, Tambov 392000, Russian Federation

[1]E-mail: pershin.home@mail.ru

Abstract. The paper shows significant influence of mode and geometric parameters on the accuracy of particles feeding process. The model of the transformation of individual material batches into a continuous flow has been proposed based on the mathematical apparatus of random Markov processes with discrete space and time. The implementation and visualization of the mathematical model is carried out through the MAPLE universal mathematical package.

1. Introduction

Virtual modeling becomes an integral part of modern computer design, since it allows studying a large number of ways for compound machine designs and constructions. This approach makes it possible to simulate real conditions of their functioning on the computer and choose the best ones for developing more perfect products [1]. Besides, it reduces time and money consumption due to the substitution of many real processes with virtual ones. In [2] the properties of the adequacy of virtual models and the conditions associated with them for the appearance of the substitution effect are described. The authors developed an approach putting virtual computer models in an intermediate position between mathematical models and simulated physical objects. They considered virtual model as mathematical and physical one realized with the help of a physical device (i.e., computer). Moreover, a new fundamental property of virtual models, the ability to partially or completely substitute physical reality by computer models as adequacy and rudeness, can be added as key criteria of mathematical models. In this regard, the possibility of a superior substitution (i.e. ensuring the conditions not yet realized to fundamentally new solutions of real physical issues) seems to be of special interest. Among the others, processing bulk materials (loose, granular and, powder materials) comprising transportation, grinding, dosing, blending, classification, etc. are associated with particles movement. Meanwhile, the motion of a particle can be considered as the motion of a rigid body based on classical mechanics laws. Due to simultaneous movement of numerous particles it is almost impossible to count their collisions, even by advanced computers. That is why, various assumptions have been employed, where the bulk material was considered as a liquid or highly concentrated gas based on the laws of hydrodynamics or aerodynamics, respectively [3].

Continuous weighing of particles is the key operation of many technological processes. The main indicator of feeders is the feeding accuracy, which characterizes the uniformity of bulk material continuous flow at the outlet of a feeder and the deviation of actual productivity from the specified one.

The accuracy of agro-food particles feeding by vibrating feeders was considered in [4]. Feed-rate deviations caused by hopper refills widely used in the loss-in-weigh industrial technology was described in [5]. The general strategy feeding accuracy increase with this technology was presented by Coperion K-Tron [6]. Besides, feeding devices improvement was studied in [7-10] .To calculate the main parameters of the feeder, various mathematical models have been developed [11-13] and feeding accuracy has been increased.

The accuracy of nanoparticles feeding is of particular importance, since their content in the final product is tenth, and sometimes, even hundredths of a percent. For example, when modifying plastic lubricants with hybrid nanomaterials (graphene and carbon nanotubes), the content of these nanomaterials is less than 0.1%. Thus, even at capacities of tens of kilograms per hour for lubrication, the nanomaterials feeding productivity should be of the order of hundredths of a gram per second.

Weighing the material during its movement is the main reason limiting the increase in the continuous feeding accuracy for bulk material. Since dynamic loads on the weight sensor reduce its accuracy, a two-stage continuous feeding technology has been proposed [14]. The essence of the technology is that separate bulk material batches of weight ΔP are formed at the first stage of the process; at equal intervals of time ΔT, they are supplied into a cylindrical tray vibrating torsionally [15]. The feeding capacity Q can be evaluated as follows:

$$Q = \Delta P/\Delta T. \quad (1)$$

Considering the aforementioned, the aim of the present work is to develop a virtual model for a two-stage continuous feeding process, which would allow determining conditions and geometric parameters which have significant impact on the feeding accuracy.

2. Modeling of particles feeding process

We choose two-stage continuous weighing of bulk materials as modeling object and the device shown in Figure 1. A cylindrical tray (2) is installed on a base (1) which vibrates due to a vibrator (3). The frequency and amplitude of oscillations are controlled by unit No. 1 (5). Individual bulk material batches of weight P are formed by a batcher (4) and are supplied into the tray (2) through a funnel (8) at regular intervals T. The batcher (4) is controlled by unit No. 2 (6) and a synchronizer (7). From the tray (2), the material is continuously supplied onto the inclined plate of a flow sensor (10), and then into a receiving container (9) or on a conveyor belt (not shown).The mathematical apparatus of random Markov processes is widely implemented for modeling bulk solid mixing [16], grinding and classification [17].

The process was considered to be discrete in space and time, so the tray and the conveyor belt along the length were divided into sections, and the transition of the system from one state to another occurs in a jump-like manner. The system was understood to be a bulk material distributed over the tray sections and the conveyor belt cells.

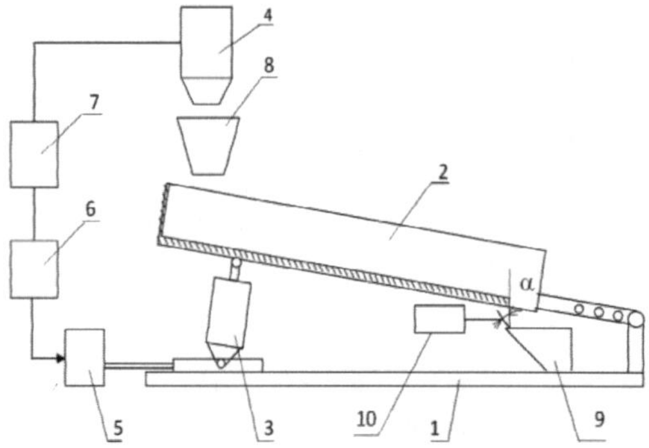

Figure 1. Two-step particles feeding: 1- base; 2- cylindrical tray; 3- vibrator; 4- batcher; 5- control unit No. 1; 6- control unit No. 2; 7- synchronizer; 8- funnel; 9- container; 10- flow sensor.

Figure 2 shows the Markov chain for the task under consideration. The first *N1* cells simulate the vibrating tray, whereas the following *N2* - the conveyor belt.

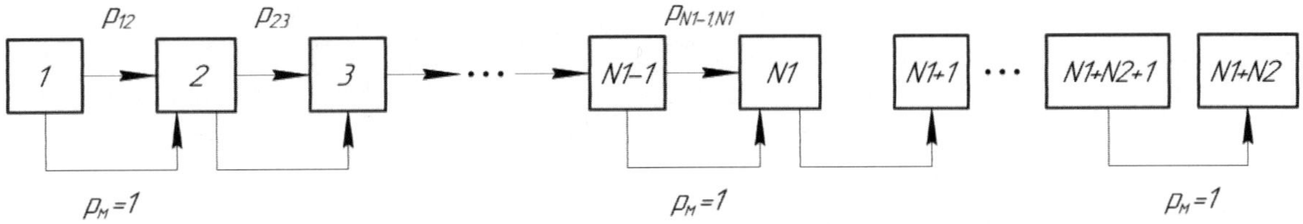

Figure 2. Markov chain of the tray and the conveyor belt.

Two processes take place on the conveyor: cell redistribution of particles and movement of the particles along the vibrating tray. The intensity of the particle redistribution is determined by the probabilities $p(i, i+1)$, and particles movement along the tray – by the probabilities $p_M(i, i+1)$.

The state of the system is described by the $S(i)$ vector:

$$S(i) = \{s(1), s(2), s(3) .... s(j), ... s(N)\}, \qquad (2)$$

where i is transition number, j – cell number, and N–the total amount of cells:

$$N = N1 + N2, \qquad (3)$$

where$N1$ is the number of tray sections or cells, $N2$ is the number of conveyor belt sections (figure 2).

Thus, *N1* elements of the state vector characterize the distribution of particles along the tray sections, whereas *N2* elements– along the conveyor cells. Depending on the task, the numerical values of $s(j)$ elements can be expressed in grams, cubic cantimeters, or individual batch weight fractions. The continuous feeding process is described by the following relationships:

$$\begin{aligned} S(1) &= S(0) \times P \times P_M \\ S(2) &= S(1) \times P \times P_M \\ &\dots \\ S(i) &= S(i-1) \times P \times P_M \\ &\dots \\ S(n) &= S(n-1) + SL \\ S(n) &= S(n) \times P \times P_M \\ &\dots \\ S(k) &= S(k-1) \times P \times P_M \end{aligned} \qquad (4)$$

where $S(i)$ is the vector of the system state after transition i, P is the matrix of transition probabilities, P_M is the movement matrix, n is the number of material batch loading steps (transitions), $n=Lm$, $m=1,2,3,\dots$(integers), L is the number of batch loading steps; $k=T_F/\Delta T$, where k is the number of transitions, after which the process ends, and T_F is the process ending time, ΔT is the time of one transition. Matrices that remain unaltered at changing transition number are generally used. The matrix of transition probabilities is given as follows:

$$P=\begin{vmatrix} 0.5 & 0.5 & 0 & 0 & 0 & 0 & 0 & 0 & 0 & 0 \\ 0 & 0.5 & 0.5 & 0 & 0 & 0 & 0 & 0 & 0 & 0 \\ 0 & 0 & 0.5 & 0.5 & 0 & 0 & 0 & 0 & 0 & 0 \\ 0 & 0 & 0 & 0.5 & 0.5 & 0 & 0 & 0 & 0 & 0 \\ 0 & 0 & 0 & 0 & 0.5 & 0.5 & 0 & 0 & 0 & 0 \\ 0 & 0 & 0 & 0 & 0 & 1 & 0 & 0 & 0 & 0 \\ 0 & 0 & 0 & 0 & 0 & 0 & 1 & 0 & 0 & 0 \\ 0 & 0 & 0 & 0 & 0 & 0 & 0 & 1 & 0 & 0 \\ 0 & 0 & 0 & 0 & 0 & 0 & 0 & 0 & 1 & 0 \\ 0 & 0 & 0 & 0 & 0 & 0 & 0 & 0 & 0 & 1 \end{vmatrix} \qquad (5)$$

In matrix (5) first five elements simulate the vibrating tray, and the rest ones model the conveyor belt. The numerical values of the matrix elements simulating the vibrating tray depend on the mode parameters (tray slope angle, frequency and amplitude of tray torsional vibrations), as well as the friction coefficients of the material being weighed (external friction coefficient, internal friction coefficient, movement), and are determined when identifying the parameters of a mathematical model for a real process(equal values of 0.5 were chosen herein).

Since in the real implementation of the two-step feeding technology of bulk material batches are supplied into the tray periodically at time intervals ΔT, the vector of the system state must periodically change in the mathematical model.

For a transition, the material can come from cell i not only to neighboring cell ($i+1$) but also to cell ($i+2$). If during one transition the batch material is redistributed among several cells, the conversion of individual batches on the vibrating tray is carried out faster. At the same time, according to calculations, the material batch leaves the tray more rapidly. In all cases, transition time and of tray sections number, maybe easily calculated from an experiment.

The elements related to the conveyor are always constant (all the elements on the first diagonal located above the main diagonal of the matrix are equal to 1).

The movement matrix has the following form:

$$P_m=\begin{vmatrix} 0 & 1 & 0 & 0 & 0 & 0 & 0 & 0 & 0 & 0 \\ 0 & 0 & 1 & 0 & 0 & 0 & 0 & 0 & 0 & 0 \\ 0 & 0 & 0 & 1 & 0 & 0 & 0 & 0 & 0 & 0 \\ 0 & 0 & 0 & 0 & 1 & 0 & 0 & 0 & 0 & 0 \\ 0 & 0 & 0 & 0 & 0 & 1 & 0 & 0 & 0 & 0 \\ 0 & 0 & 0 & 0 & 0 & 0 & 1 & 0 & 0 & 0 \\ 0 & 0 & 0 & 0 & 0 & 0 & 0 & 1 & 0 & 0 \\ 0 & 0 & 0 & 0 & 0 & 0 & 0 & 0 & 1 & 0 \\ 0 & 0 & 0 & 0 & 0 & 0 & 0 & 0 & 0 & 1 \\ 0 & 0 & 0 & 0 & 0 & 0 & 0 & 0 & 0 & 1 \end{vmatrix} \qquad (6)$$

IOP Conf. Series: Journal of Physics: Conf. Series **1084** (2018) 012005 doi:10.1088/1742-6596/1084/1/012005

At each transition, a multiple of $\Delta T/\Delta t$, where Δt is the time of one transition, the following procedure should be performed:

$$S(0) = S(0) + S_L, \tag{7}$$

where S_L is the vector, which elements characterize particles distribution shape along tray sections at loading.

In fact, the considered mathematical model allows simulating the fluctuation of batch shape for particles moving along the vibrating tray. Besides, it is possible to transform individual batches into a continuous flow with a periodic distribution change of next batch in initial tray sections.

In the virtual model, the batcher' functioning, i.e. the process of forming individual material batches is proposed to be simulated using a random number generator and a filter providing the required deviations of the individual batch weight from a given value. The range of numerical deviation values depends on the particular batcher used in two-stepfeeding. This process (i.e., individual batch feeding into the vibrating tray), as well as the periodic redistribution of the material in the initial sections is performed by employing specially developed software.

In the present work, the cell distribution of the batch after its loading into the cylindrical tray was calculated using the T-Flex CAD software. The MAPLE universal mathematical package (Waterloo Maple Corporation, Waterloo, ON, Canada) was chosen herein as a modelling tool for developing the model. This software can be used for numerical and symbolic calculations; it enables simulation of technical systems and visualization of a wide variety of calculations and physical processes. Besides, it provides tools for preparing technical documentation. Moreover, the software contains a unit for estimating root mean square deviations of the material weight in the sampler cell from the required values based on the given capacity. This unit makes it possible to use the virtual program to determine the accuracy of continuous weighing depending on the mode and geometric parameters of the device, on which the two-stage feeding technology is implemented.

3. Results and discussion

Figure 3 shows the distribution of the material along the sections: cells 1 to 19 simulate the vibrating tray, and cells 20 to 50 – the conveyor belt. Figures 3a-3c presents options where the next material batch is loaded every 30, 15 and 10 steps, respectively. In all the histograms, the material is distributed directly after loading the next batch. Initially, this material is distributed only along cells 1 and 2, and then, under the action of vibration, it is distributed along the rest tray cells. The ratio of the weight of the material located on cell i to the weight of the individual batch represents specific gravity. Using a relative weight enables scaling any performance, without changing the vector of the system state $S(0)$ and the matrix of transition probabilities P.

As can be seen from the histograms, the distribution of the material among the sections when supplied at the 10th (Figure 3c) or the 15th (Figure 3b) steps is more uniform than when the material is supplied at the 30th step (Figure 3a). It should be noted that the actual weight of the individual material batch shown in Figure 3a is twice larger compared to Figure 3b, or three times more compared to Figure 3c, since the feeding process with the same capacity is simulated.

At first glance, it may seem that the easiest way to increase the uniformity of the continuous flow of the bulk material at the outlet of the dispenser is to supply the individual batches into the vibrating tray more often (i.e., to reduce the T value). However, in this case, the weight of one batch will decrease, and consequently, weighing errors will increase. We assume that the batcher weighs the individual batches with an accuracy of 0.001 g. If the batch weight is 1 g, then the error will be 0.1 %, but if it is 0.5 g, then the error will be 0.2 %. Naturally, increasing the error of weighing the individual batch will increase the error of the continuous feeding.

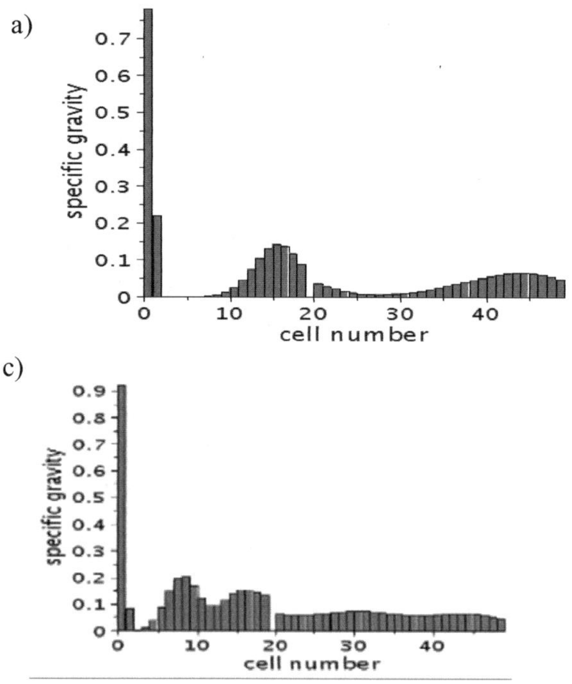

Figure 3. Histograms of the distribution of bulk material along the sections of the tray and conveyor: the material is supplied at the 30 (a), 15 (b), and 10 (c) steps.

Figure 4 shows the distribution of the material immediately after loading the third batch. It can be seen that the third portion was distributed on the first two sections of the vibrating tray. The forms of the previous (first and second) batches changed, and they merged into a continuous flow.

The calculation of the movement of the individual material batch along the inclined vibrating tray using the model described above yields the similar results.

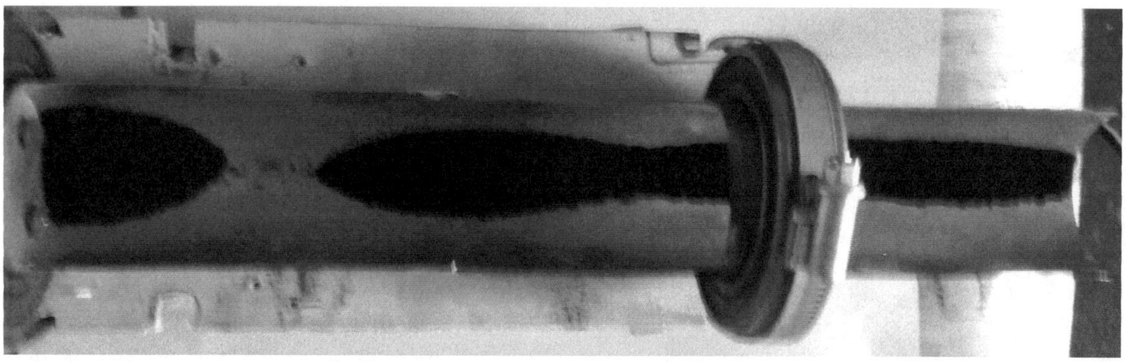

Figure 4. Distribution of the material immediately after loading the third batch into the vibrating tray.

The next batch is loaded prior to complete discharge of the current batch from the tray (here next batch is loaded at 30 steps). Since the "stretching" batch is faster than its discharging from the tray, the adjacent batches are interconnected. Therefore, the individual material batches are converted into a continuous flow. In a real feeder, the variable parameters are as follows: tray slope angle, amplitude of torsional vibrations, and weight of the individual batch weight.

To identify the parameters of the proposed mathematical model for a specific bulk material, it is sufficient to carry out nine simple experiments with three different amplitudes of torsional vibrations and three different tray slope angles. In these experiments, it is necessary to determine the batch distribution immediately after loading into the tray, time when the particles are discharged from the tray, time during which all particles are discharged from the tray, and (4) conveyor belt cell distribution of the particles.

The adequacy of the model to the real process is achieved by changing the elements of the matrix of transition probabilities. The optimization objective is the minimum deviation from a given performance at any specified time.

Conclusion

The virtual model of the two-stage continuous weight feeding of bulk materials presented herein allows using the minimum of experimental data to estimate mode and geometric parameters of the vibrating feeder, which provides the specified performance and accuracy. Besides, the model enables tracing the physical evolution of the feeing process, monitoring the planned sequence of bulk material batch conversion into a continuous flow, and visualizing the geometric details of each device component.

Thus, the proposed model can be used in the design of new devices for continuous weight feeding of bulk materials with high accuracy, even at low performance (less than 0.1 g s^{-1}). Moreover, this model can be applied for educational purposes, since it allows elucidation of the type and operation method for equipment involved in the two-step feeding of bulk materials.

Acknowledgments

The work was performed within the framework of the Federal Target Program "Research and Development in Priority Areas of the Scientific and Technological Complex of Russia for 2014-2020" (State Contract No. 14.577.21.0253, 2017; Unique Identifier for the Applied Scientific Research: RFMEFI57717X0253).

References

[1] Sampaio A Z and Henriques P G 2008 Virtual reality technology applied on the visual simulation of construction activities. *Open Construction and Build. Technol. J.* **2** 7
[2] Makarov I M, Rakhmankulov V Z and Akhrem A A 2007 Virtual modeling and intelligent management of complex computer-integrated systems. *Inf. Technol. Computer Systems* **2** 11
[3] Bonamy D, Chavanis P H, Cortet P P, Daviaud F, Dubrulle B and Renouf M 2009 Euler-like modelling of dense granular flows: application to a rotating drum *Eur. Phys. J. B* **68** 619
[4] Ola D C and Popescu S 2009 Considerations regarding the accuracy of dosing equipments for agro-foods bulk solids that use vibratory feeding systems. *Bull. Transilvania University of Braşov.* **2** (**51**) Series II 99
[5] Engisch W E and Muzzio F J 2015 Feedrate deviations caused by hopper refill of loss-in-weight feeders. *Powder Technol.* **283** 389
[6] Nowak S 2017 Three ways to improve continuous loss-in-weight feeding accuracy. Aavailable at: https://www.coperion.com/media/4317/2017_en_pbei_improving-liw-feeding-accuracy.pdf
[7] Kozak A 2016 Mathematical model of elements of automated system of loose materials dosing. *Ukraine journal of mechanical engineering and materials science* **2**/**1** 65
[8] Bukreev V G, Lyapushkin S V and Gusev N V 2014 Accuracy enhancement of electrotechnical system for bulk material bathing. *MATEC Web of Conf.* **19** 01025
[9] Bukreev V G, Lyapushkin S V and Edgulov M M 2015 Accuracy enhancement of automated system for multicomponent batching of bulk materials. *MATEC Web of Conf.* **37** 01012
[10] Demchenko V A and Kazakov Y R 2015 Ways to improve the accuracy of dispensers for bulk foods. *Sci. J. NIU ITMO* **1** 37 [in Russian]
[11] Gopireddy S R, Hildebrandt C and Urbanetz N A 2016 Numerical simulation of powder flow in a pharmaceutical tablet press lab-scale gravity feeder. *Powder Technol.* **302** 309
[12] Hou Q, Dong K J and Yu A 2014 DEM study of the flow of cohesive particles in a screw

IOP Conf. Series: Journal of Physics: Conf. Series **1084** (2018) 012005 doi:10.1088/1742-6596/1084/1/012005

feeder. *Powder Technol.* **256** 529
[13] Abhishek B P, Patel V J and Agrawal P M 2012 Solid Modeling and Analysis of Vibrating Grizzly Feeder (VGF). *Int. J. Eng. Res. Develop.* **1(8)** 34
[14] Pershin V F and Baryshnikova S V, RU Patent No. 2138784 (02 June 1998)
[15] Alsayad T K and Pershin V F 2017 Movement of the bulk material portionon on cylindrical tray that makes torsion oscilation *XXIV Int. Conf. 'Mechanical Engineering and Technosphere'* 16, September 8-17, 2017, Sevastopol, Russia
[16] Mizonov V, Balagurov I, Berthiaux H and Gatumel C A 2017 Markov chain model of mixing kinetics for ternary mixture of dissimilar particulate solids. *Particuology* **31** 80
[17] Mizonov V E, Berthiaux H, Zhukov V P and Bernotat S 2004 Application of multi-dimensional Markov chains to model kinetics of grinding with internal classification. *Int. J. Miner. Process* **74** S307

IOP Conf. Series: Journal of Physics: Conf. Series **1084** (2018) 012006 doi:10.1088/1742-6596/1084/1/012006

Methods of virtual modeling of dynamic educational programs in terms different manufacturing sectors requirements

V Martynov [1], E Filosova [1], O Shiryaev [1] and P Sakal [2]

[1] Institute of Economics and Management, Ufa state aviation technical university, 12 K. Marx Street, Ufa, Republic of Bashkortostan, 450008, Russian Federation
[2] Faculty of Materials Science and Technology in Trnava, Slovak University of Technology in Bratislava, Vazovova 5, 812 43, Bratislava, Slovakia

[1]E-mail: vvmartynov@bk.ru
[2]E-mail: peter.sakal@stuba.sk

Abstract. The article considers problems connected with the professional standards implementation at an enterprise. The principles of dynamic construction of educational programs are revealed. The approach to the formation of dynamic educational programs taking into account the requirements of the professional community and foreign experience is proposed. To cut the transition costs we suggest that a specialist selection process as well as dynamic design and methodological support formation should be automated for the implementation of the new generation curricula according to the professional standards requirements. This approach can be applied, among other things, to the training of physicists for various production sectors.

1. Introduction

Professional standards implementation in many countries has been successful in terms of labour efficiency growth. The report [1] provides an analysis of different methods of developing and using professional standards in Germany in comparison with other countries. The interaction between the education system and the requirements of the German labour market is achieved through the application of the integrated principle of the development of vocational training provisions as an educational standard that includes a professional standard. In Germany, as elsewhere in the world, there is a close link between professional standards and vocational and theoretical programmes in the development of curricula and didactic principles. It should be noted that training is not carried out on the specific requirements of the employer, so the employees has to get additional education. There is also no possibility to adapt the curricula in the process of their implementation to the requirements of the professional community. In Russia, the Unified Rating and Skills Guides for jobs and occupations of manual workers and the Reference Book of managers, specialists and workforce qualification are getting outdated: they either contain no new occupations or their descriptions don’t line up with reality. The Ministry of Labour and Social Protection of the Russian Federation has signed up and published a lot of professional standards to create the All-Russian qualifications framework for all economic sectors. About 60 standards have been adopted in mechanical engineering by now and more than 40 of them are coming up for discussion. According to the Federal Act of May 2, 2015, No. 122-FZ the employers have been required to use them since June 1, 2016. The project realization will allow us to solve the problem of demand for skilled personnel in mechanical engineering by 2018. Based on the obtained result, we can say that the examination of welders' professional competence has made it possible to reduce accident incidence rate by 20-25 per cent at "Gazprom" and "Rosneft" facilities. Besides, the new legislative acts are designed to coordinate educational and professional standards and eventually eliminate the disconnect between the learning process and the professional association as well as standardize the requirements for the skills level and job functions with the learning outcomes.

2. Methods

The notion of "professional standards" in the given regulatory documents is defined as a qualification characteristic required for an employee to perform a certain kind of professional activity (part 2, art191.1 of the Labour Code of the Russian Federation). In the professional standards the description of requirements for a specialist includes more contemporary schemes such as combination of knowledge, skills, occupational skills and work experience. From the beginning of the academic year, the training and retraining of specialists will start in all educational establishments according to these and subsequent standards. But the professional standard is more connected with reality as it is based on the hands-on experience of the specialist's occupational activity rather than educational programs.
In the near future, the development of new Federal State Educational Standards (FSES) 3++ will begin. They will contain the characteristics of the graduate's professional activity, universal competencies (uniform to the level of education in all areas), general professional competencies, that set general requirements for the area of study and are independent of the focus of training. So the problem that has arisen in recent years consists in the fact that a graduate of an educational institution has certain professional skills, while the employer needs quite different ones. The given problem is especially important for the field of mechanical engineering as it is a high technology and quickly developing branch, after all it accounts for more than one third of overall commodity output of Russian industry.

Organizations of various sectors of the economy try to employ graduates of higher educational institutions with well-formed professional qualities that allow them to provide higher labour productivity, with the appropriate competencies that will make it possible not to increase the company's non-production expenses on additional training and retraining of personnel. If the employee does not comply with the professional standard, then, according to the Article 196 of the Labour Code, the employer has to organize training, which will require additional funds. It is obvious, that in such situation the employer will prefer to employ an employee who does not need additional training or it should be minimal.

To this end, at the moment when forming the Federal State Educational Standards for vocational education, it is necessary to take into account the provisions of the relevant professional standards, but within a year they will have to be updated and brought into compliance.

The enormous work carried out to develop and introduce new educational standards is designed to reduce the distance between employers and educational institutions and create a more effective system for their interaction. The main areas of cooperation between the university and employers are not only work on the modernization and implementation of educational standards and programs; they also include the establishment of basic departments for large employers, cooperation with them in the professional standards formation, taking steps towards advancing through professional and public accreditation of educational programs.

In Russia, machine building industry in the Council for Vocational Qualifications in Machine Building is represented by All-Russian industrial association of employers "Russian Engineering Union". There are changes in the requirements for the workforce: uniformity and interchangeability of workers give way to the "uniformalised" type of employee; blurred boundaries between professions are formed, under which identification of the traditional types of labour is lost; the isolation of professional castes is destroyed, and professions are "globalized".

The same set of functions and qualifications of a specialist can be a competitive advantage in the labour market at a specific time, and at another time it is of no importance for a specialist to be in demand.

When developing and implementing higher education programs, educational institutions should first and foremost focus on the needs of employers, which will create a mechanism for continuously monitoring changes in the labour market conditions and the requirements of key consumers for the quality of education. Therefore, practice of employers' participation in the educational activities of institutions (Figure 1) and in the evaluation of quality of education is becoming commonplace in Russia, because employers need specialists of the required qualifications who are able to start working without additional training.

For employers, the task consists in forming professional standards in the light of which the Federal State Educational Standards are designed. The duration of standards formation is supposed to take up to three years. Nevertheless, even if the new FSESs are released within a year (which is unlikely), it will be necessary in accordance with them to develop curricula for the graduation of new specialists. Thus, it will take 8-10 years before the first cadre of students will graduate from university educated according to the FSESs taking into account the requirements of employers. In fact, nowadays specialists have the competencies that were required about 10 years ago and even earlier.

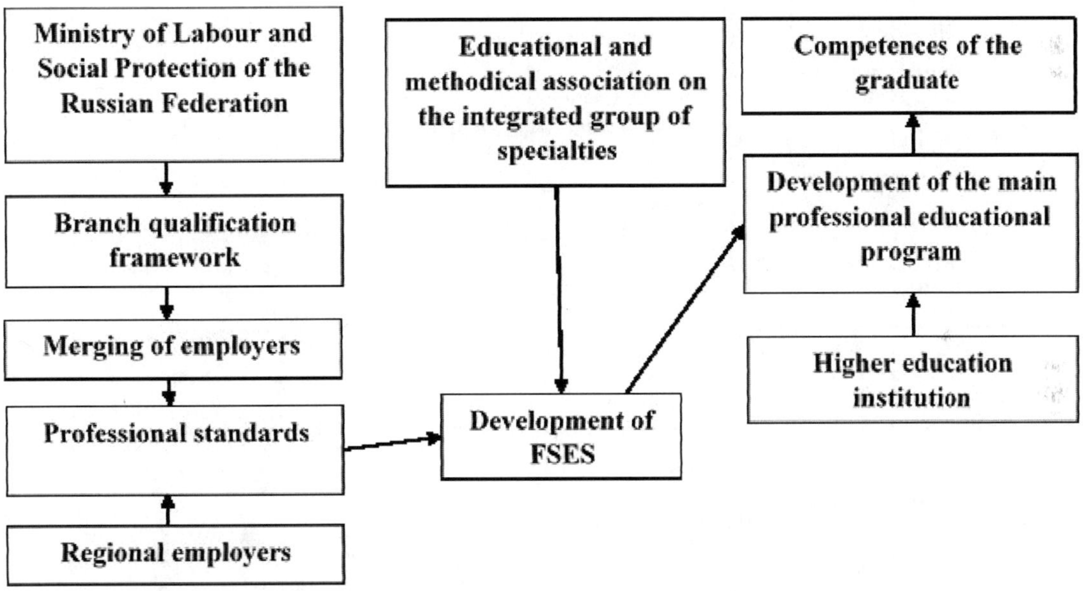

Figure 1. Professional standards influence on the development of educational programs.

A natural question arises as to how to accelerate this process as much as possible. You can try to predict the competencies that will be required after 10-15 years and provide advanced training, but it does not mean that the forecast will be confirmed. There is also a problem of rapidly changing requirements for the graduates of higher professional education for the developing branches of the national economy.

Therefore, a task of operational training of a certain number of students having the required competences is becoming now more urgent.

This task can be solved with the help of the formation of dynamically changing educational programs that reflect the relationship between the theory and its definite practical application, corresponding to the vocational aptitudes of the future bachelor, specialist or master [2].

Such dynamic educational plans can also be useful in upgrading the skills of employees working in various sectors of the economy, so that they could get the right qualifications.

Within the framework of the standard approach, it is rather difficult to describe the elements of the educational activity and the connections between them. The federal state educational standard of higher professional education considers competencies to be an ability to apply knowledge, skills and personal qualities for successful activities in a certain field.

In general, the formation of agreed requirements for the competencies of a specialist includes the following stages:

- building a business model of production processes;
- determination of specialist's core competences;
- harmonization of requirements with the FSES;
- analysis and formation of the hierarchical structure of requirements;
- approval of a set of requirements by the parties involved;
- formation of the training sequence in the form of a training program;
- object-oriented analysis of the training programs feasibility.

Figure 2 shows the harmonizing of the requirements for the professional standard with the FSES within the competency-based approach.

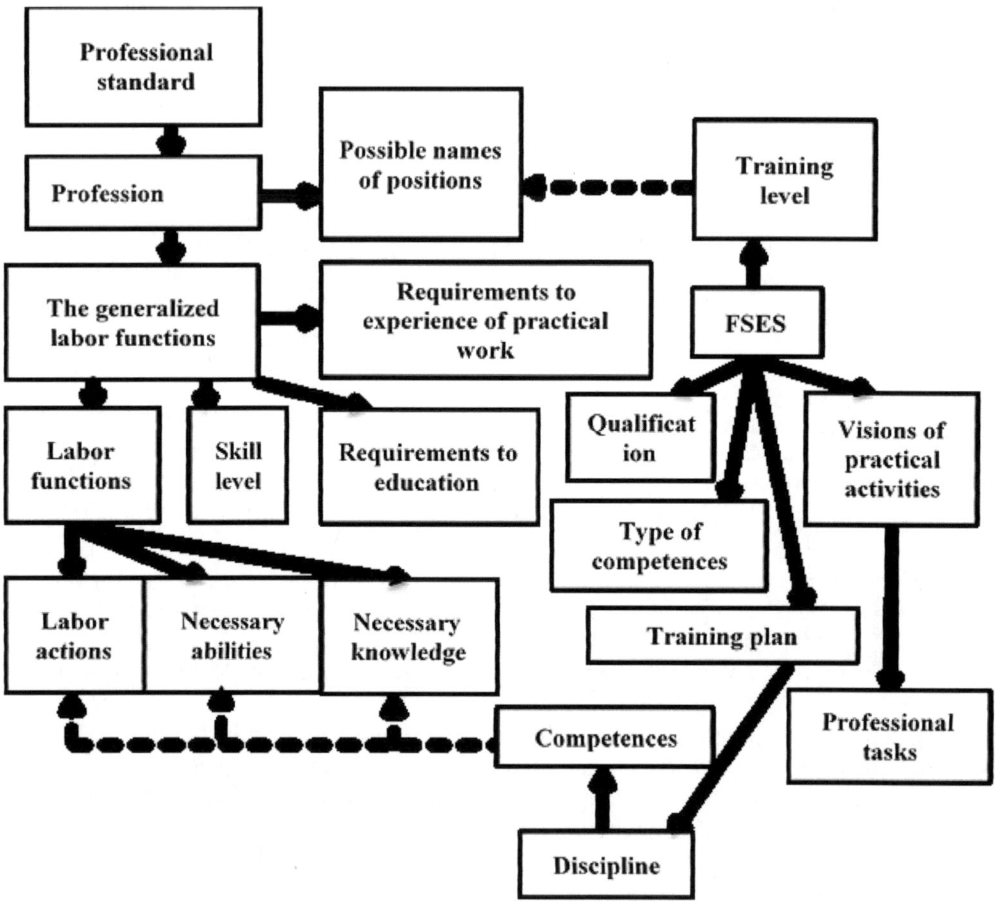

Figure 2. Federal State Educational Standard (FSES) and professional standard interaction.

As noted earlier, the experience of Germany as well other foreign countries is analyzed by *W. Petersen* and *M. Jepsen* in report [1], in which the approach to the development of structure and professional profile for new "professional standards" and "educational professions" is proposed. This approach is based on the change of professional profiles, focused on the theory and scientific and technical knowledge, to professional profiles, focused on production and business processes. Thus, the new "professional standards" and "training professions" are changing in a direction based on "competence or qualification" and "orientation towards learning outcomes". The main innovation of these changes is that three descriptors of "knowledge, skills and competencies" should be used for all qualification levels of education and training.

In accordance with the above-described approach [1], the development of educational standards adopted in Russia is close to foreign.

3. Results

The prerequisites for establishing a learning management system of new generation, which should satisfy the need in a changing business environment and education, are described in [3-5]. From this and in accordance with the analysis of other sources it follows that the method of formation of dynamic educational programs with the requirements of employers is new.

One of the problems arising in the selection of specialists by the employer is the discrepancy between the qualification requirements given in the professional standards and the competencies that are presented in the training standards. Therefore, the first task in reaching harmonization is preparing a thesaurus for the subject area, constructing and using a mechanism of determining the necessary competencies. Such thesaurus can be formed as an ontological database and later general requirements can be developed based on the built-on ontology [6].

For example, let's consider the ontological model of the master degree educational process developed in compliance with the requirements of employers that was implemented in the Protege 4.2 software product (Figure 3) [7, 8]. This ontology allows the developers of educational programs and employers to obtain information that meets the criteria of concern [9]. So, the employer can get information about the competencies formed by graduate students in the process of preparation and will be able to formulate the requirements for the prospective employee on the basis of the received information analysis.

In addition, this ontology makes it possible to determine elective disciplines for master students in accordance with the employers' requirements. For example, an employer wants a potential employee to perform a work function of "Managing the identification and implementation of IT innovations". This work function is included under the professional competence of "Ability to conduct searches and analyse innovations in the economy and management" from the educational standard of the "Business Informatics" training area (38.04.05).

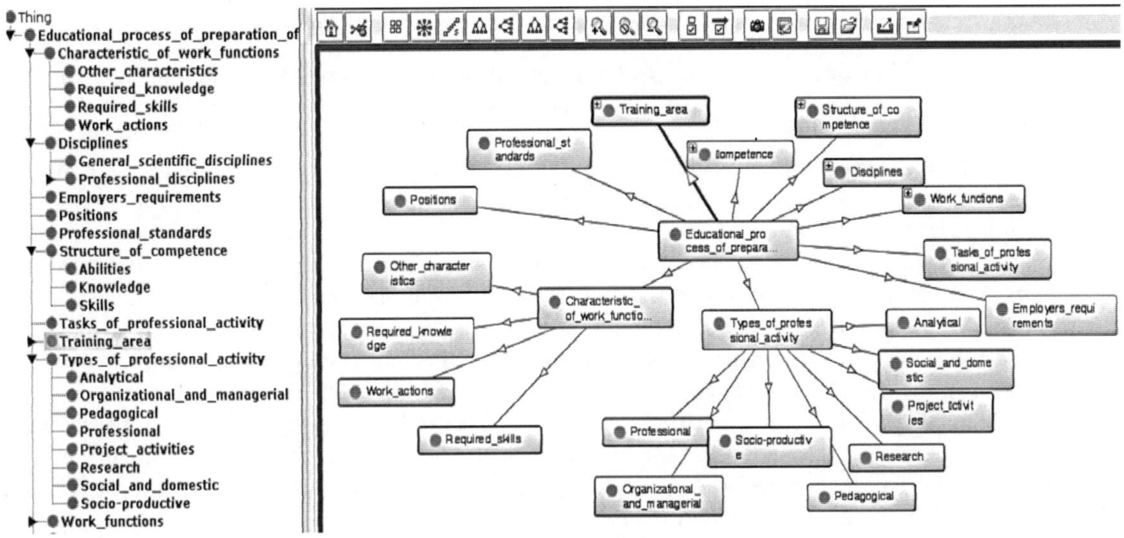

Figure 3. A fragment of ontology of the educational process of master degree students training.

This competence can be formed with the help of elective discipline "Information technologies in the development of managerial decisions" (Figure 4). Thus, the ontology allows determining the elective disciplines for the master degree training in accordance with the requirements of employers and, thereby is used to form an educational program that meets the requirements of employers.

As a result of ontological modelling, the appropriate knowledge base was created for specialists in the field of information technologies that includes both the requirements of the FSES and formalized requirements of professional standards. The labour functions needed for a certain position were put in line with professional tasks. The side effect of the detailed description of the user's competencies is that it dramatically simplifies the employer's task of selecting the necessary personnel.

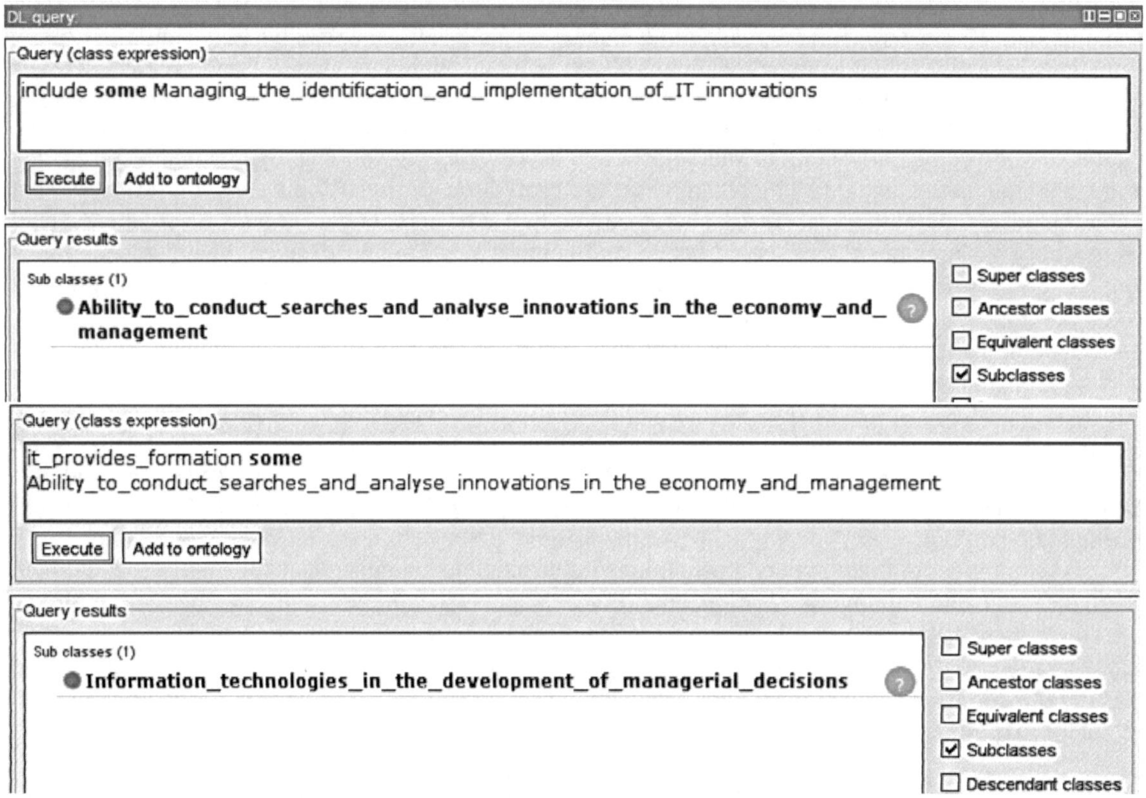

Figure 4. An example of finding a discipline that meets the requirements of the employer in terms of the professional standard.

Several hundred competencies of different levels from different areas of knowledge that can describe the model of a modern graduate's preparedness provide an effective formalized search for candidates of the right profile and necessary qualifications.

For the realization of this task the informational system was created where the employer's requirements formation proceeds in several stages. At first, the employer chooses a profession from the professional standards, if there is none, the employer inputs the profession himself. Description of occupations can be found in the Directory of Occupations in the reference guide section. Further, the employer determines the required competencies based on the professional standard. These competencies can be further determined by the university. At the final stage, additional competencies related to the requirements of a particular employer are added. As a result of the query concerning the given competencies, it is determined the federal educational standard that is closest to the developed model in terms of the structure and content of the basic competencies.

Based on the specified requirements the most proper standard is selected. The employer gets a list with the most appropriate training areas and the level of the required competencies satisfaction that can't be found in the existing FSES (Figure 5). These competencies are formed within the variable part of the curriculum or in the course of additional training if the variable part cannot cover the competence difference. A specialist's training area that meets a set of the required competencies and differs little from the competencies formed on the basis of the training standard, is determined based on the interplay of a variety of the specialist's competencies and a host of the FSES competencies.

Training direction	Percentage of realization	Unfulfilled competence
230100	27.272727272727 %	The ability to carry out selection of the source data for the design of The ability to conduct modeling of processes and systems The ability to carry out the calculation of health and safety conditions Willingness to participate in the work on the final design and development of information technologies in the course of implementation and operation of information systems The ability to use technology development facilities of professional activity in different areas The ability to participate in the formulation and conduct of experimental studies The ability to execute operating results obtained in the form of presentations , scientific and technical reports , articles and papers on the cientific and technical conferences The willingness to ensure the security and integrity of information systems and data technology
230400	23.529411764706 %	Develop components of software systems and databases , using modern tools and programming techniques Participate in setting up and commissioning of hardware-software complexes Install hardware and software for information and automated systems

Figure 5. Unsatisfied competences.

To design or change the curriculum of the training area closest to the competencies that are being formed, qualification requirements are decomposed to the level of competencies and subsequently to the level of disciplines. On the basis of the performed decomposition, a compliance matrix of the competencies and constituent parts of the educational program (disciplines, training units) is designed, on the base of which a curriculum in this area is subsequently developed. Based on the description of the competencies presented to a specialist that are not formed in the chosen area of study, the designing of changeable or replaceable disciplines is carried out. After that it is necessary to develop training and methodological support for teaching the disciplines to be added or changed.

Training toolkits are developed by the teacher on the base of the knowledge and experience he or she possesses, as well as materials from libraries or the Internet. On the basis of the educational standard the teacher compiles a list of issues that are obligatory for studying in this discipline, and forms a list of modes of education based on the curriculum and the complexity of the discipline. After that, the issues are divided into lectures in accordance with their amount. Search for materials on the theme of the lecture as well as synopses and lectures preparation are carried out using the materials provided by the teacher and/or information available from the library and/or the Internet.

After putting together the lectures, materials are prepared for conducting other types of academic activities: laboratory operations manuals, methodological guidelines for practical training, course papers, or computational and graphic work. This allows automatic formation of the necessary methodological support for the discipline from the invariant parts in accordance with the requirements of the educational standards.

IOP Conf. Series: Journal of Physics: Conf. Series **1084** (2018) 012006 doi:10.1088/1742-6596/1084/1/012006

4. Conclusion

With the application of the competency-based approach to the design and development of the new generation educational programs and using the object-oriented approach to the structuring of knowledge, it becomes possible to dynamically update educational programs in accordance with the requirements of the new professional standards. A data model used for knowledge representation has a decisive effect on the functionalities of computer-aided data processing application in any subject area. The advantage of ontologies as a way of representing knowledge is their formal structure that simplifies their computer-aided processing.

Dynamic design and provision of the teaching and methodological support for the educational programs implementation may be used to address the issue of increasing the training effectiveness of the in-demand specialists due to a deep analysis of the requirements made by all the entities interested in their competitiveness (university-government-employer), and through the study process and curriculum building automation.

This paper is also part of the project VEGA No.1/0235/17 "System identification of complex assumptions to support industrial innovation and employment in the less developed regions of Slovakia", built on the results of successfully finished project KEGA No. 037STU-4/2012 "Implementation of the subject of "Sustainable corporate social responsibility" into the study program of Industrial Management in the second degree of study at STU MTF Trnava" and APVV No. LPP-0384-09 "Concept of HCS model 3E vs. concept Corporate Social Responsibility".

References

[1] Petersen W and Jepsen M 2015 *Report "German approach and experience of development of occupational standards"* (Flensburg) p 47 http://www.ilo.org/wcmsp5/groups/public/---europe/---ro-geneva/---sro-moscow/documents/publication/wcms_397644.pdf

[2] Martynov V V, Filosova E I and Filosova V K 2016 Automating the formation of dynamic training plans for FSES 3+ based on the requirements of the employer *Proc. Int. Conf. ITRT-2016 (Togliatti)* (Russia) 54-61

[3] Stone D E and Zheng G 2014 Learning Management Systems in a Changing Environment *Handbook of Research on Education and Technology in a Changing Society* **2** pp 756-767

[4] Ryneveld L V 2016 Introducing Educational Technology into the Higher Education Environment: A Professional Development Framework *In Innovative Professional Development Methods and Strategies for STEM Education* **8** 126-136

[5] Kelly M, Costello M, Nicholson G and O'Connor J 2016 A Collaborative Academia-Industry Approach to Developing a Higher Education Programme in Building Information Modelling *International Journal of 3-D Information Modeling (IJ3DIM)* **5(2)** 39-54

[6] Hitzler P, Gangemi A, Janowicz K, Krishnadi A and Presutti V 2016 *Ontology Engineering with Ontology Design Patterns: Foundations and Applications* (IOS Press) p 388

[7] OWL 2 Web Ontology Language Quick Reference Guide (Second Edition) Available from https://www.w3.org/TR/2012/REC-owl2-quick-reference-20121211/

[8] Guarino N 1996 *Understanding, Building, and Using Ontologies* Available from http://ksi.cpsc.ucalgary.ca/KAW/KAW96/guarino/guarino.html

[9] Ternai K and Szabo I 2016 Semantic Application for the Internationalization Audit of Higher Education Institutions *Proc. Int. Conf. EGOVIS 2016* (Portugal) 194-205

IOP Conf. Series: Journal of Physics: Conf. Series **1084** (2018) 012007 doi:10.1088/1742-6596/1084/1/012007

Modelling of spindle nodes for machining centers

O Krol[*] and V Sokolov

Machinery Engineering and Applied Mechanics Department, Volodymyr Dahl East Ukrainian National University, 59a Central Avenue, Severodonetsk, 93000, Ukraine

[*]E-mail: krolos@yandex.ru

Abstract. Unified spindle nodes constructions of machining centres for drilling-milling and boring types are considered in this paper. 3D modelling of the main components for the spindle units on the rolling bearings is performed, including the construction of complex assemblies, kinematic schemes for shaping the body parts and various kinds of tool modules for machining. The efficiency of transparency mapping procedure both in realistic and grid mode is noted. Parametric models of spindle nodes are developed on the basis of dyplexed angular contact ball bearings, established according to the schemes "dyplex" and "triplex". The geometric core of the APM Graph module is used to parametrically set the position of the rod elements and the coordinates of their nodes in the 3D-representations of the machine spindle. Two feedback loops have been introduced in the complementary procedure for spindle modeling related to the change in the types of supports and calculation methods. A study of the efficiency to design of the spindle-boring block assembly by the rigidity criterion is carried out.

1. Introduction

Metal processing has been industrially applied for drilling-milling and boring machines that have significantly increased technological capabilities due to the integration of processing centers into complex-automated systems [1-3]. These machining centers are distinguished by module design principle of their construction and possibility to create numerous machine tool modifications. Meanwhile, the effectiveness of its introduction is largely related to the development of forming units using the modular principle. Growing diversity of tool blocks nomenclature (as auxiliary and cutting tools) and rising requirements top-product quality need complex studies of spindle nodes for machining centers in Computer Aided Design (CAD) environment [4].

In numerous studies of workability associated with stress-strain state of spindle nodes (ShN), the initial stage of research efficiency is creating 3D model of these nodes in CAD environment. Based on 3D model, new approach for the estimation of static stiffness in machine working space has been analyzed by means of high-performance computerized Computer Numerical Control (CNC) [5], which considers a parametric model with six directional static rigidity for designing and evaluating rigidity in machine working space. Products machinability during its life cycle of a triaxial milling machine equipped with tool rigging of type Bridgeport DIN 69871 are considered in [6]. The authors use system approach to estimate processing efficiency based on the interaction between basic forming units of machine tool like feed drive, spindle unit and tool system by author. Application of ISO 230 series as auxiliary tool corresponds to machines of first and second sizes. CAD SOLIDWORKS and CAD ANSYS software was fully used for building 3D model of a spindle with a CNC lathe and its finite element method [7]. When the spindle is formalized, a 10-node rod model with three degrees of freedom is used in each node. Analysis of dynamic characteristics at 5 natural frequencies and spindle oscillation modes with allowance for changing load made it possible to predict the behavior of the product outside the resonance zone and to estimate the rigidity index with the domination of bending stresses. At the same time, the authors noted the need to simplify the original 3D model (removing bevels, small holes and other structural elements) in the process of converting 3D file formats between SOLIDWORKS and ANSYS.

For effective design and research of the forming units for the machining center, the corresponding sections of the parameterized databases of spindle nodes structures and specialized tooling are needed. To enhance the competitiveness of such 3D models for construction must be supplemented by a photorealistic representation using rendering and animation programs. Hence the requirements for resources and capabilities of applied CAD. The variety of technological operations and the corresponding design schemes leads to the need to take into account feedback in the process of creating construct. The choice of support bearings (especially the front spindle supports) and loading methods predetermines the position and coordinates of the rod elements and their nodes, and thus the shape and dimensions of the 3D model. Moreover, the calculation methods and procedures of the investigation will depend on the loading scheme, as well as the number of iterations in the mode of multivariate design for the spindle units.

This article is devoted to the creation of sets of 3D models for spindle nodes of the most common third-level and fourth-dimensional machining centers and modeling of their designs. The main software product is integrated CAD/CAM/CAE KOMPAS-3D with built-in modules Artisan Rendering and APM FEM finite element analysis of structures.

The article deals with the features of the process for designing two-bearing construction of spindle assemblies, which are mounted on rolling bearings with different layouts and various preload forces. For varied machining operations, the applied tool blocks (for drilling, milling and boring), introduce changes in the procedures for constructing models of the structures spindle unit and the methods for calculating them.

2. 3D modelling of spindle nodes

Modern forming units of metal-cutting machines are closed dynamic systems using modular tooling equipment [8]. In order to examine the stress-strain state and touse finite element methods it is necessary to build 3D models of individual parts and assemblies of ShN in the integrated CAD KOMPAS-3D with the APM FEM module [9-11].

As a basic construct, the ShN of CNC machining centers the MTs200, SF68, SF16 and SBM1 models [12], embedded in the main drive of the machine tools is considered. The developed 3D model of drive for the SF68VF4 machine with the use of KOMPAS-3D resources specifying the parametric links and connection between 3D assembly individual components is presented in Figure 1.

The drive under consideration in the matching mode with the turntable provides machining of the parts from all sides, as well as coaxial boring of the holes without reinstalling workpieces. The kinematic chains realizing the execution of a complex for technological operations typical for drilling-milling and boring machines are shown in Figure 2a and 2b.

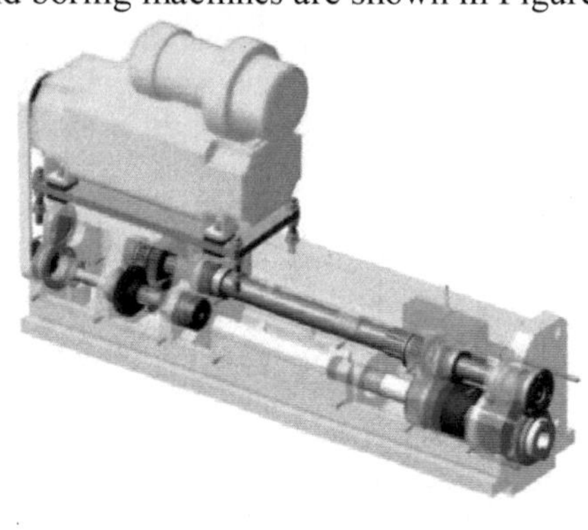

Figure 1. Two-stage drive of the main movement of machine tool SF68VF4 model.

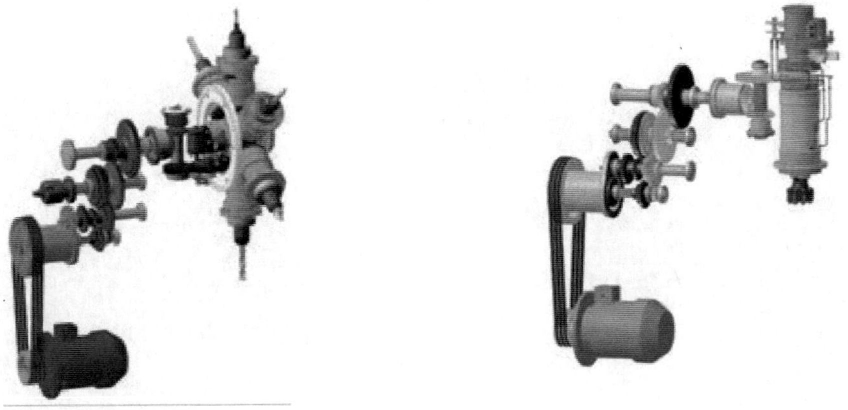

a b

Figure 2. 3D kinematic schemes of machine tool SF16MF3 (a) and SVM1VF4 (b).

3. Controlling the image of the spindle assembly

To create ShN 3D models we used image control operation related to the transparency display gaining modeling procedure productivity. Although computer resources were not fully used, fast loading of complex assemblies and fast rotation were carried out.

The grid and realistic transparency mapping was also used in assembly editing mode (these settings were applied only to objects that were not edited). Nevertheless, there were difficulties in realistic transparency design (Figure 3a) due to errors in displaying objects behind semi-transparent object. Therefore, we used mesh transparency promoting correct mapping of overlapped objects. Figure 3b shows ShN 3D model for the machining center MTs200VF4 developed by the authors in the mode of grid transparency.

a b

Figure 3. Effects for ShN in 3D modeling.

To provide information support and the promotion of machine tool production, it was necessary to represent product using attractive and informative images of its form and design. Artisan Rendering application is generally used since it is integrated into CAD KOMPAS-3D CAD and allows obtaining high-quality image and design documentation [12]. So, we applied Artisan Rendering settings to obtain photorealistic view of spindle nodes of MTs200VF4 machines (Figure 4a) and SF16MF3

(Figure 4b) based on the control of lighting parameters, camera characteristics, and background selection.

a b

Figure 4. Spindle node rendering: a – MTs200VF4; b – SF16MF3.

It is often required to evaluate the mutual motion of ShN parts and to monitor the trajectories to identify design limitations. Creation of animation rollers for projected ShN helps designers and technologists to quickly understand the arrangement of the product and to learn assembly-disassembly order. For SF16MF3 machines (Figure 5a) and SVM1VF4 (Figure 5b), assembling - disassembly animations have been built.

a b

Figure 5. Animation tools in CAD KOMPAS-3D: a – SF16MF3; b – SVM1VF4.

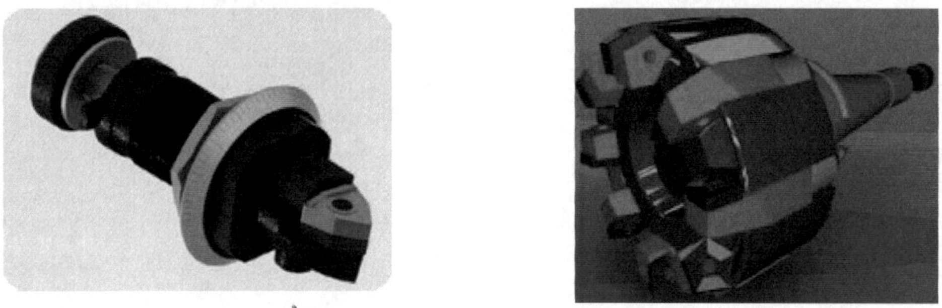

a b

Figure 6. Tool blocks of machining centers: a – boring tool; b – milling tool.

Along with spindle nodes unification, we consolidated and tooled machining centers. Figure 6 shows the design of the tool blocks used in the above-mentioned CNC machine tools – boring for hole machining (Figure 6a) and milling for processing flat surfaces (Figure 6b).

4. Discussion

When constructing ShN 3D model for CNC machining centers of drilling-milling-boring type equipped with a variety of tooling, there is strong attention to the relationship between tooling' type and design and constructive features of ShN models [4]. This refers both to spindle schemes and to 3D models based on the rod elements. In the process of constructing the rod element on the stage of determining its length, the construct of the front support was taken into account. Here rod node position was determined by type, number, and installation scheme of bearings. Three-rod representations connecting the drive element assemblies (gear [13], toothed half coupling [14] and belt transmission [13] – coordinate *b*) are used for spindle nodes for machining center; the support nodes are the coordinate *l* and the spindle offset from the front support is the coordinate *a*. The high-precision radial-thrust two-row tapered roller bearing [15] is installed on the front support of the spindle in the machine SF16MF3 [15], providing increased stiffness of the support (Figure 7a).The node's position of the rod 2-3 is determined by coordinate *a* (Figure 7b). The front support of triplex-X type of MTs200VF4 machine, is mounted in the dyplexed radial-thrust bearings (Figure 8a) and the node of the rod 2-3 is determined by the coordinate *a*.

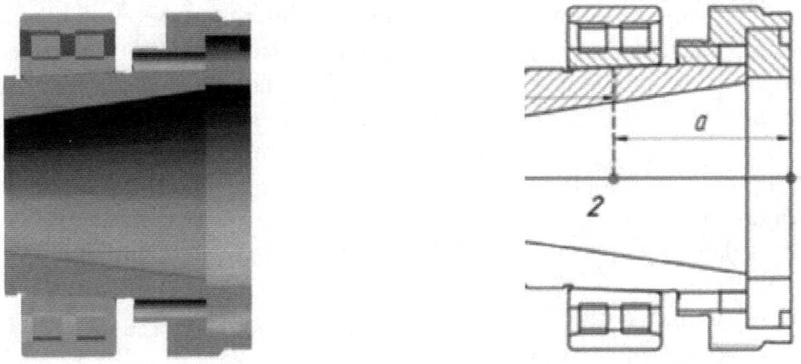

a b

Figure 7. Front bearing on double row roller bearing: a – 3D construct; b – constructive scheme.

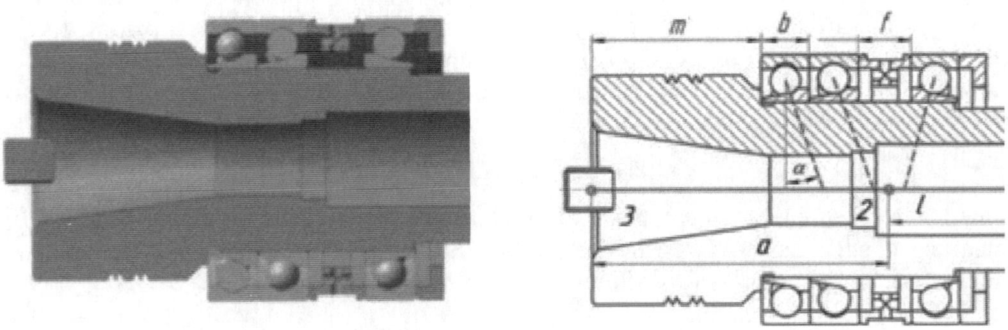

a b

Figure 8. Front support on double angular contact bearings: a – 3D-construct; b – constructive scheme.

Depending on support type, the coordinates of rod solid model for the spindle unit have been determined, and in particular the initial coordinate of the third rod (rod 2-3, Figure 8 and Figure 9) simulating the spindle cantilever part in the point of tool block fixation. This problem is effectively solved at the parametric level [16-18] in the APM Graph module, a fragment of the variable window for determining the coordinate *a* is shown in Figure 9.

sgl	1*(N>=1&N<2)+0*(N>=2&N<	0
dup	1*(N>=2&N<3)+0*(N>=3&N<	0
Tri	1*(N>=3&N<4)+0*(N>=4&N<	1
x1	(0.5*d+0.25*(Dp-d))*tan(rad(.	20.1189
a	(m+b/2+x1)*sgl+(m+b+b/2+>	106.881

Figure 9. Fragment program of parametric model (window of variables).

To the position of the rod nodes is also affected by the preloading force in bearings mounted as "Duplex" and "Triplex" [19]. There are empirical dependencies associated with the change in the contact angle α from the value of the preloading force in radial-thrust bearings. The developed program for parametric determination of the coordinates of the nodes of the 3D model for the spindle takes into account the pre-loading change factor. This makes it possible to specify the points of application of the applied load and to obtain a more real picture of the displacement and stress fields in the spindles.

In the process of research of the projected construction, the condition for single-valued fixation in space should be specified, and also check for the presence of the rod nodes with allowed displacement under the forces. To represent the spindle unit in the form of a beam on two hinged supports, with prohibited displacement in the direction of the external force, such free nodes are absent. In this case, you have to enter an additional free node in which it is allowed move on certain coordinates. As an example of such a node it is possible to bring a node of a rod simulating the effect of a drive element - a toothed half-coupling (Figure 4a) [14]. Thus, there is a feedback between the stage of building the 3D model and the stage of the research construct operability, for example, by the criterion of rigidity.

Another aspect is related to the choice of the method for solving problems in evaluating the construct rigidity [20]. Thus, for tool holders working on bending with an insignificant torque (this refers to the operation of thin boring), it is sufficient to use the initial parameters in the power calculation. Otherwise, with a significant fraction of the torque (for end milling), the finite element method is used. In this case, the cross-section of the tool holders (in the form of a rod) is divided into flat finite elements that interact with each other by means of nodes. This situation leads to feedback (the second type) between the stage of building the 3D model and the stage of choosing the calculation method of the projected construction [21]. For the thin boring operation, the method of initial parameters in the matrix formulation [12] was used and static forms for estimating the rigidity (compliance) were obtained. For MTs200VF4 machining center for the case of a standardized ShN with installed boring holder, the compliance δ will be defined as a function of the length holder l_k according to the analytical dependence

$$\delta = (3395+31.47 l_k+0.13\, l_k^2)\ 10^{-8},\ \text{mm/N}$$

This method can serve as an alternative one to the finite element method to provide simplified and fast evaluation of stiffness characteristics of unified ShN designs with a changing tooling that is typical for the machining centers of a drilling-milling-boring group.

Conclusion

A complex research of spindle units for machining centers of drilling-milling-boring type in the environment of integrated CAD KOMPAS-3D and APM Win Machine, equipped with a variety of instrumental block for processing complex body parts was implemented. 3D-models of the elements for the main motion drive with spindle forming units were developed using the parameterization apparatus in the APM Graph module and rendering in the Artisan Rendering module.On the basis of these modules the workability research is carried out and the competitiveness of the machines created increases. Three-dimensional solid models of spindle nodes for various construct and their kinematic schemes are presented. The algorithms for parameterization of construct schemes on the basis of rod elements are developed, their position is determined depending on the bearing type in the support, the method of mounting and the value of the preload. This makes the process of designing spindle nodes in the multivariate mode more efficient. Investigations of the ShN according to the rigidity criterion are carried out, and analytical relations to evaluating the compliance of unified two-bearing ShN equipped with a variety of tooling is obtained

The proposed approach to structures modeling allows a complex procedure for research spindle nodes of machining centers are implemented. This procedure on the basis of developed parametric models, drawings and 3D models in the mode of multivariate design is completed.

References

[1] Abrahamova T M, Bushuyev V V and Gilova L Ya 2012 *Handbook on metal-cutting machine tools* [in Russian] (Moscow: Mechanical Engineering) p 608

[2] Balmont V B, Gorelik I G, and Figatner A M 1987 *Calculations of high-speed spindle node*s [in Russian] (Moscow: VNIITEMR) p 52

[3] Push AV and Zverev I A 2000 *Spindle nodes. Designing and research* [in Russian] (Moscow: Stankin Publishing) p 197

[4] Grigoriev S N, Grechishnikov V A and Maslov A R 2012 *Instrumental systems of integrated machine-building productions* [in Russian] (Moscow: Stankin Publishing) p 194

[5] Gao X, Li B, Hong Ju and Guo Ju 2016 Stiffness modeling of machine tools based on machining space analysis *J. Adv. Manufact. Technol.* **86/5-8** 2093

[6] Afsharizand B, Zhang X, Newman S T and Nassehi A 2014 Determination of machinability *Proc. 47th CIRP Conf. on manufacturing systems*, **17** 760

[7] Kong J and Cheng X 2017 Modal analysis of CNC lathe's spindle based on finite element. *J. Advances in Engineering Research (AER)* **148** 318

[8] Pronikov A S, Borisov E I and Bushuev V V 1995 *Handbook Designing of metal-cutting machine tools and machine-tool systems* [in Russian] (Moscow: Mechanical Engineering Publishing) p 371

[9] Zamriy A A 2013 *Textbook Practical training course CAD/CAE APM Win Machine* [in Russian] (Moscow: APM Publishing) p 144

[10] Ganin N B 2012 *Three-dimensional design in KOMPAS-3D* [in Russian –Trehmernoe proektirovanie v KOMPAS-3D] (Moscow: DMK Publishing) p 776

[11] Shelofast V V and Chugunova T B 2004 *Fundamentals of machine design. Examples of solving problems* [in Russian] (Moscow: APM Publishing) p 472

[12] Krol O S, Shevchenko S V and Sokolov V I 2011 *Textbook Design machine tools in environment APM WinMachine* [in Russian] (Lugansk: V. Dahl EUNU Publishing) p 386

[13] Reshetov D N 1989 *Details of machines* [in Russian] (Moscow: Mechanical Engineering Publishing) p 496

[14] Shevchenko S, Mukhovaty A and Krol O 2017 Gear clutch with modified tooth profiles *J. Proc. Eng.* **206** 979

[15] Shevchenko S, Mukhovaty A and Krol O 2016 Geometric aspects of the modified tapered roller bearings *J. Procedia Eng.* **150** 1107

[16] Girschtick J 2018 Introducing parametric modeling 2.0. *J. Isicad* **162** 8

[17] Rosinsky S, Shanin D and Grigoriev S 2001 Parametric capabilities of the graphic module of the APM Graph of the APM Win Machine system *J. CAD and Graphics* **11** 37

[18] Nemtinov V, Bolshakov N and Nemtinova Y 2017 Automation of the of the early stages of plating lines design *MATEC Web of Conf.* **129** 01009.

[19] Balmont V V 1990 Statics and dynamics of high-precision ball bearing units [in Russian] *Proceedings of Moscow State University: Machine Building Series* **155**

[20] Sokolov V and Krol O 2017 Installations criterion of deceleration device in volumetric hydraulic drive *J. Procedia Eng.* **206** 936

[21] Nemtinov V A, Nemtinova Y V, Borisenko A V and Nemtinov K V 2014 Construction of concentration fields of elements in 3D in groundwater of an industrial hub using GIS technologies *J. Geochem.Explor.* **147** 46

IOP Conf. Series: Journal of Physics: Conf. Series **1084** (2018) 012008 doi:10.1088/1742-6596/1084/1/012008

Mathematical modeling of the stress-strain state of flexible threads with regard to plastic deformations

D Tarasov [1], V Konovalov [2], V Zaitsev [3] and Y Rodionov[4]

[1]TseSIS NIKIRET, 62 Chaadayev street, Penza, 440067, Russian Federation
[2]Faculty of Industrial Technologies, Penza state technological University, 11 Gagarin street, Penza, 440015, Russian Federation
[3]Faculty of Industrial Technologies, Penza state technological University, 11 Gagarin street, Penza, 440015, Russian Federation
[4]Tambov State Technical University, 106 Soviet street, Tambov, 392000, Russian Federation

[2]E-mail: konovalov-penza@rambler.ru

Abstract. The object of the study is the main power elements of mechanical systems made of flexible threads working on the perception of transverse loads. The paper focuses on the calculating the stress-strain state of a flexible thread with nonlinear elasticity by means of MathCAD program based upon geometric and physical nonlinearity of flexible threads. The proposed method is based on mathematical modeling theory including differential and integral calculation of one and several variables functions, as well as on the points of materials strength. We propose a method for nonlinear calculation of the stress-strain state of flexible threads taking into account the real properties which may be introduced to study the work of structures outside the elastic work of the material. Comparison of the parameters of the stress-strain state of a flexible thread obtained with the help of a digital experiment in a computer program, also with the results obtained by the finite element method, confirmed the adequacy of the proposed method and the method of mathematical modeling. Divergences in the values of the calculated parameters were no more than 5%.

1. Introduction

Cables and ropes are widely used in industry, construction, transport and other manufacturing sectors. At the same time these structures are flexible elements, and their behavior, deformation and trajectory significantly affect the performance of these [1, 2].

Well-known theories and hypotheses as impact for rope systems [1], finite element method [3, 4] and classical calculations [5-8] used for establishing spatial relationships and types of loading for the calculation of flexible elements allows one to calculate parameters of these structures with sufficient accuracy. The calculation of flexible threads has been performed for a different external effect. Meanwhile, the validity of such hypotheses has been confirmed by long-term service of operating structures, carried out in full-scale and model experiments. However, there is still no reliable and accurate theory describing the behavior of the rope as a flexible element [1, 2, 9] with nonlinear properties of flexible threads.

The purpose of the work is to develop a method for calculating the stresses and strains of flexible threads with nonlinear elastic properties, to check the adequacy of such mathematical model with the help of a computer program developed on the basis of functional expressions that have the form close to theoretical dependencies.

2. Research methods

The object of research was a horizontal flexible thread with a lateral load. Based on the known information [9] and taking into account a number of additions, the stress-strain state of flexible threads was described and a computer program for their calculation and optimization was developed.

In modeling the stress-strain state (SSS) of flexible threads, the following assumptions were accepted: the material continuously filled the entire volume (the continuity hypothesis); the properties of the material were the same at all points (homogeneity) and were independent of direction (isotropy); the transverse normal sections remained flat after deformation (Bernoulli hypothesis); the transverse section was small in comparison with the length; the change in transverse section upon stretching was not taken into account; from the internal forces there was only tension (axial force), and rigidity was absent in bending and torsion.

The assumptions made it possible to replace the flexible thread with an equivalent model – a flat curve – its centerline. When compiling the calculation model, the category of the formation of a rod type element was used [2]. If the height of the transverse section of the rod was negligible compared to its deflection, then such a rod was called flexible. Rods can perceive various loads, but among them, a load of their own weight was required. The evenly distributed load acted over the entire length of the rod. Under the action of its own weight, the flexible rod assumed an equilibrium state in the form of a certain curve.

3. Research results

An elementary design model – a rod on two hinged supports located on the same level was considered. Since the rod was flexible and could not perceive bending moments, the forces arising from the external action in any section of it including the supports were directed only along the rod itself. Therefore, the reactions of the supports were directed along the tangents to the sag curve. Designated they were denoted as T_A and T_B respectively. The origin was chosen on the left support. The coordinate axes were directed as shown in Figure 1 [9].

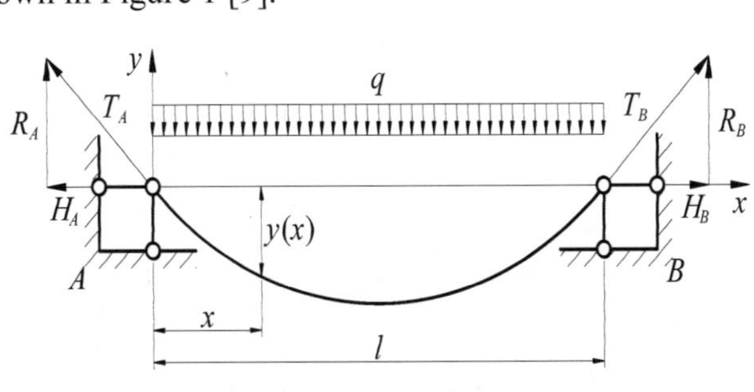

Figure 1. Flexible thread design model: T_A, T_B – tension of the thread points A and B; R_A, R_B – vertical reaction of the support at point A and B; H_A, H_B – horizontal reaction of the support at points A and B; q – distributed lateral load; l – distance between supports; x –horizontal coordinate of the desired point; y – vertical coordinate of thread desired point (difference in the heights of the desired point and A point); y(x) –function of the thread deflection coordinate.

Each of the support reactions was decomposed into two components: vertical and horizontal. To determine these components, we used the equations of statics. First of all, by projecting all the acting forces on the horizontal axis, we established the following equality:

$$H_A = H_B = H. \tag{1}$$

IOP Conf. Series: Journal of Physics: Conf. Series **1084** (2018) 012008 doi:10.1088/1742-6596/1084/1/012008

The quantity H was henceforth called the flexible thread axial force; at points A and B H was designated as H_A and H_B.

Then, equating to zero the sum of the bending moments of all acting forces with respect to some reference point, for example A, we obtained reaction at point B as R_B:

$$R_B = \frac{\sum M_A}{l}, \tag{2}$$

where $\sum M_A$ –sum of the bending moments of all external forces with respect to point A, with the exception of the reference reactions, H×m; l – span, m.

Of the support reactions components, only R_B created a bending moment, since the remaining forces passed through point A. In the particular case of a uniformly distributed load and of supports located on the same level, we had the reaction of the supports at points A and B, respectively:

$$R_A = R_B = q \cdot l \cdot \frac{l}{2} \cdot \frac{1}{l} = \frac{q \cdot l}{2}, \tag{3}$$

where q – uniformly distributed load, N/m.

It could be seen from formula (3) that the determined vertical component was equal to the reaction that would had arisen in the hinged single-span beam by the same span and loaded just like the considered flexible rod. Since the thread was flexible, the sum of the bending moments of all forces acting on it on one side of any unspecified cross section taken at a distance X from the support A was also zero:

$$\sum M_x - H \cdot y(x) = 0, \tag{4}$$

where $\sum M_x$ –sum of the bending moments of all vertical forces, including the vertical component of the support reaction, located to the left of the section with abscissa x, H×m; $y(x)$ –ordinate of the considered section, m.

The total bending moment (4) corresponded to the bending moment of the single-span beam, loaded as the flexible rod. The magnitude of the total moment for the middle of the span would be:

$$\sum M_x = \frac{q \cdot l}{2} \cdot \frac{l}{2} - q \cdot \frac{l}{2} \cdot \frac{l}{4} = \frac{q \cdot l^2}{8}. \tag{5}$$

The expression obtained (5) is a well-known formula for determining the maximum bending moment in the middle of a span by a hinged single-span beam loaded with a uniformly distributed load. Such a moment was called a beam point.

To determine the ordinate of an unspecified cross-section along the length of a flexible thread, we rewrote formula (4) in the form:

$$y(x) = \frac{M(x)}{H}. \tag{6}$$

After obtaining the fundamental dependence (6), it could be concluded that the deformed state of the flexible thread under the action of the vertical load coincided with the bending moment diagram of the hinged beam by the same span under the same load. In this case, the ordinates of the moment diagram were reduced by dividing by the axial force value and were laid off from the chord AB connecting the attachment points.

From expression (6) it could be seen that the construction of the equilibrium line reduced to determining axial force. Since the determination of the bending moment function in a hinged single-span beam did not cause difficulties.

Since the spread was not known until the time of determination, and it entered equations for calculating the initial and final length of the flexible thread, all subsequent expressions were written as functions of the axial force:

IOP Conf. Series: Journal of Physics: Conf. Series **1084** (2018) 012008 doi:10.1088/1742-6596/1084/1/012008

$$L_0 + \Delta L(H_1) + \Delta L_t = L_1(H_1)\,, \tag{7}$$

where L_0 – initial length, m; $\Delta L(H_1)$ – elastic deformation, m; ΔL_t – temperature deformation, m; $L_1(H_1)$ – final length, m.

However, it was no longer possible to determine the elongation of a flexible thread when plastic deformations were taken into account by formula (8) based on Hooke's law.

$$\Delta L(H_1) = \frac{H_1 - H_0}{E \cdot A} \cdot \int_0^l \left[1 + \left(\frac{Q_1(R_{1A}, R_{1B}, x)}{H_1} + \tan\beta \right)^2 \right] dx\,, \tag{8}$$

where H_0, H_1 –axial force caused by the initial load and the combined action of the initial and additional load, N; E –modulus of elasticity of the material, Pa; A – cross-sectional area, m^2; $Q1(R_{1A},R_{1B},x)$ – transverse force function in the hinged beam by span l from the combined action of the initial and additional load, N; x – current abscissa ($0 \le x \le l$), m; β –slope angle of the chord AB, deg.

To determine the elongation of a flexible thread, it was necessary to find the relative deformation on the deformation diagram by knowing the stresses. And from the relative deformation found, the desired elongation could be determined.

Any kinds of diagrams could act as those of the material state. They determined the relationship between stresses and relative deformations. For example, curvilinear or piecewise-linear diagrams. Most importantly, they had to respond to the true behavior of the material. It was required to specify the main parametric points of the diagram. Such points were the maximum stresses and the corresponding relative deformations, boundary values, and so on.

We took a piecewise linear deformation diagram as a working diagram of the state of the material. With its help, you can approximate almost any curve. It is presented in Figure 2.

With such a diagram, the tensile stresses, depending on the relative deformations of the elongation of the material, were determined by the formula:

$$\sigma = \begin{cases} E \cdot \varepsilon, \ \varepsilon < \varepsilon_1; \\ \sigma_1 + E_1 \cdot (\varepsilon - \varepsilon_1), \ \varepsilon_1 \le \varepsilon < \varepsilon_2; \\ \dots\dots\dots\dots\dots\dots\dots\dots\dots\dots \\ \sigma_n + E_n \cdot (\varepsilon - \varepsilon_n), \ \varepsilon_n \le \varepsilon < \varepsilon_{n+1}, \end{cases} \tag{9}$$

where E_n – hardening modulus of the material on the nth segment of the deformation diagram, Pa, ε – relative deformation of corresponding segments, fractions; σ – thread tension in the respective segments, Pa.

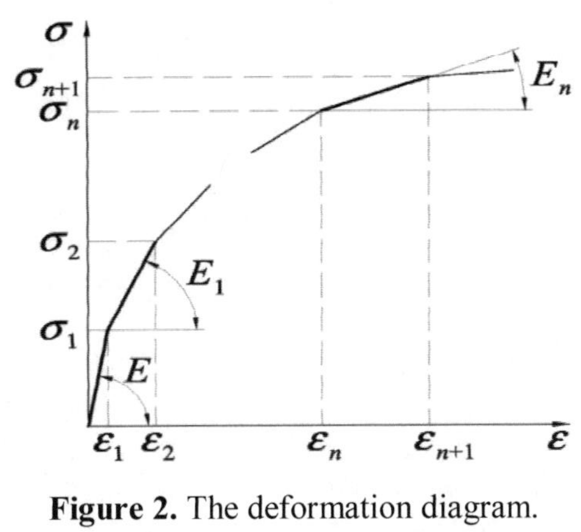

Figure 2. The deformation diagram.

The material hardening module on the nth segment of the deformation diagram could be determined from the equation:

$$E_n = \frac{\sigma_{n+1} - \sigma_n}{\varepsilon_{n+1} - \varepsilon_n}, \tag{10}$$

To find the stresses σ, we multiplied both sides of expression (8) by the value $\frac{E}{L_0}$:

$$\frac{\Delta L(H_1) \cdot E}{L_0} = \frac{(H_1 - H_0) \cdot E}{E \cdot A \cdot L_0} \cdot \int_0^l \left[1 + \left(\frac{Q_1(R_{1A}, R_{1B}, x)}{H_1} + \tan\beta \right)^2 \right] dx \tag{11}$$

After the reduction and transformation, we got tension σ:

$$\sigma(H_1) = \frac{H_1 - H_0}{A \cdot L_0} \cdot \int_0^l \left[1 + \left(\frac{Q_1(R_{1A}, R_{1B}, x)}{H_1} + \tan\beta \right)^2 \right] dx \tag{12}$$

An axial force was introduced in the parameters of the stress function, caused by the joint action of the initial and additional tension, since it was unknown at the time of the determination of the stresses.

Knowing the stresses and the fact that the relative deformation was $\varepsilon = \frac{\Delta L(H_1)}{L_0}$, we found from expression (10) the elongation of the flexible thread:

$$\Delta L(H_1) = \begin{cases} \frac{\sigma(H_1) \cdot L_0}{E}, \ \sigma(H_1) < \sigma_1; \\ L_0 \cdot \left(\frac{\sigma(H_1) - \sigma_1}{E_1} + \varepsilon_1 \right), \ \sigma_1 \le \sigma(H_1) < \sigma_2; \\ \dots\dots\dots\dots\dots\dots \\ L_0 \cdot \left(\frac{\sigma(H_1) - \sigma_n}{E_n} + \varepsilon_n \right), \ \sigma_n \le \sigma(H_1) < \sigma_{n+1}, \end{cases} \tag{13}$$

After determining the elongation caused by a change in the external force, as well as finding the initial and final length of the flexible thread according to the formulas presented in [9], it was possible to find the value of the axial force. To do this, it was necessary to solve the nonlinear equation (7) using numerical methods.

We rewrote equation (7) in the form (14):

$$f(H_1) = L_0 + \Delta L(H_1) + \Delta L_t - L_1(H_1) = 0. \tag{14}$$

We considered one of the simplest numerical methods for solving nonlinear equations – the chord method.

This method assumed that there were sufficiently small neighborhoods, each having only one root of the equation. Taking one of the points of this neighborhood as the initial approximation of the root, we could calculate the desired root with a given accuracy. From the physical meaning of the problem, such a neighborhood was the value of the axial force from zero to the value of the breaking force of a flexible thread, which depended on its cross-sectional area.

We considered the function $f(H_1)$ be continuous on the interval [a; b] and $f(a)>0$,;$f(b)<0$. To find the root of equation (14) belonging to the interval [a; b] it was necessary to divide that segment with respect to $f(a)/f(b)$. Thus, the first approximation of the root c was at the point of intersection of the interval [a; b] with a chord passing through the points with coordinates (a; $f(a)$) and (b; f (b)). Then the same method was applied to the interval [a; c] or [c; b], depending on which end the function $f(H_1)$ had opposite signs. Thus, the following approximation of the root was obtained. Graphical interpretation of the method of chords was presented in Figure 3.

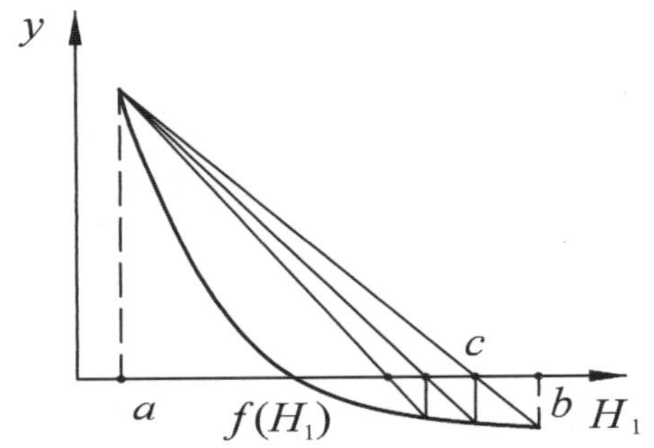

Figure 3. Graphical interpretation of the method of chords: *a,b* – coordinates of a thread segment, *c* – coordinate of a change of sign of function *y; H_1* – a thrust caused by the initial load and the combined action of the initial and additional loads.

We obtained an iterative formula for finding the approximate value of the root *c* of equation (14). We wrote the equation of a straight line passing through points with coordinates (*a*; *f*(*a*)) and (*b*; *f*(*b*)):

$$\frac{y-f(a)}{f(b)-f(a)}=\frac{c-a}{b-a}\rightarrow y=\frac{f(b)-f(a)}{b-a}\cdot(c-a)+f(a)\,. \tag{15}$$

The straight line given by equation (15) intersected the abscissa axis under the condition that *y*=0. We found the point of intersection of the chord with the abscissa:

$$\frac{f(b)-f(a)}{b-a}\cdot(c-a)+f(a)=0\rightarrow c=a-\frac{f(a)\cdot(b-a)}{f(b)-f(a)}\,. \tag{16}$$

Thus, we had:

$$c=a-\frac{f(a)}{f(b)-f(a)}\cdot(b-a)\,. \tag{17}$$

After finding the axial force, it became possible to determine the longitudinal force that occurred in the cross sections along the length of the flexible thread. The longitudinal force function was:

$$T_1(x)=\sqrt{H_1^{\,2}+(Q_1(R_{1A},R_{1B},x)+H_1\cdot\tan\beta)^2}\,. \tag{18}$$

Different but statically equivalent local loads lead to the appearance of the same stress state in the flexible thread, if local stresses were not taken into account near the point of application of the load. This state was a linear (uniaxial) stress state. It was known that with simple stretching (flexible threads can only perceive tensile forces) in all sections parallel to the axis of the element in question, the normal and tangential stresses were zero. Thus, with simple stretching at each point of the flexible thread, the main areas were perpendicular and parallel to its axis. The main stresses at these areas were equal respectively to:

$$\sigma_1(x)=\frac{T_1(x)}{A};\ \sigma_2(x)=\sigma_3(x)=0\,. \tag{19}$$

As an assessment of the adequacy of the developed method of mathematical modeling, we carried out a comparative analysis a flexible thread SSS. We defined SSS by different methods of calculation. For this we considered the finite element method (FEM) and the proposed modeling technology on the example of the scheme which design model is shown in Figure 4.

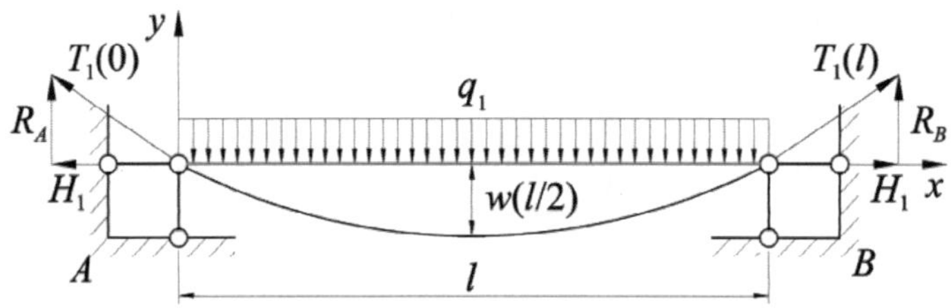

Figure 4. The design scheme of a flexible thread: — is initial state of the equilibrium line; — is final state lines of equilibrium from external influence; $T_1(0)$, $T_1(l)$ – tension of the thread at points A and B; R_A, R_B – vertical reaction of the support at points A and B; H_1 – horizontal reaction of the support at point A and B; q_1 – distributed lateral load; l – distance between supports; x –horizontal coordinate of the desired point; y – the vertical coordinate of the desired point of the thread (the difference in the heights of the desired point and the point A); $w(l/2)$ – middle deflection.

Flexible thread with supports located at the same level (span l = 6 m, cross-section $A = 0.00031416$ m^2) was loaded with a uniformly distributed load throughout the span. The initial bending deflection and the load causing the initial outline were missing. The account of the elastoplastic behavior of the material was modeled by a piecewise linear relationship between stresses and deformations. The parameters of the law of nonlinear deformation, shown in Figure 5, were expressed in the stresses of 300, 500, and 700 MPa and the corresponding relative deformations of 0.00204, 0.02, and 0.03, respectively.

We took maximum longitudinal force that occurred in cross-sections located on the supports and deflection in the middle of the span as the main criteria for comparison in determining flexible thread SSS.

To do this, we solved a number of problems, consistently equating the uniformly distributed load $q1$ to a value of 3, 9 and 20 kN/m to obtain stresses on each segment of a given deformation diagram.

The calculation of FEM was performed in the software package "LIRA-CAD" version 2015 release of R4. The construction was modeled by universal dimensioned finite rod elements with allowance for physical and geometric nonlinearity (type 410). The breakdown was made for 100 final elements. The calculation was carried out by a stepping non-linear processor, designed to solve physical and geometrical nonlinear and contact problems. At each step, the stiffness matrix was formed in the "new position" coordinate system, taking into account the change in the tangential deformation modulus.

As an example, Figure 6 and Figure 7 show the results of SSS modeling of a flexible thread loaded with a uniformly distributed load $q_1 = 3\ kN/m$. The greatest longitudinal force in the finite element method (Figure 6, Table 1) corresponds to 85.7 N. The implementation of the proposed method (Table 1) gives a value of 85.8 N. The error of the values is small and is 0.12%. Similarly, the movement of the thread (Figure 7, Table 1) on the calculation options is 0.162 m and 0.158 m, respectively for the proposed method and FEM. The error of the values is small and is 2.47%.

Table 1. Values of a flexible thread stress-strain state (SSS) evaluation criteria calculated with the help of FEM and developed simulation technology.

Evenly distributed load, kN/m	**3**		**9**		**20**	
	$T_1(0)$, $T_1(l)$, *kN*	$w(l/2)$, *m*	$T_1(0)$, $T_1(l)$, *kN*	$w(l/2)$, *m*	$T_1(0)$, $T_1(l)$, *kN*	$w(l/2)$, *m*
Proposed method	85.8	0.158	134.3	0.308	187.9	0.507
FEM	85.7	0.162	136.0	0.307	185.0	0.517
Inaccuracy, %	0.12	2.47	1.25	0.32	1.54	1.93

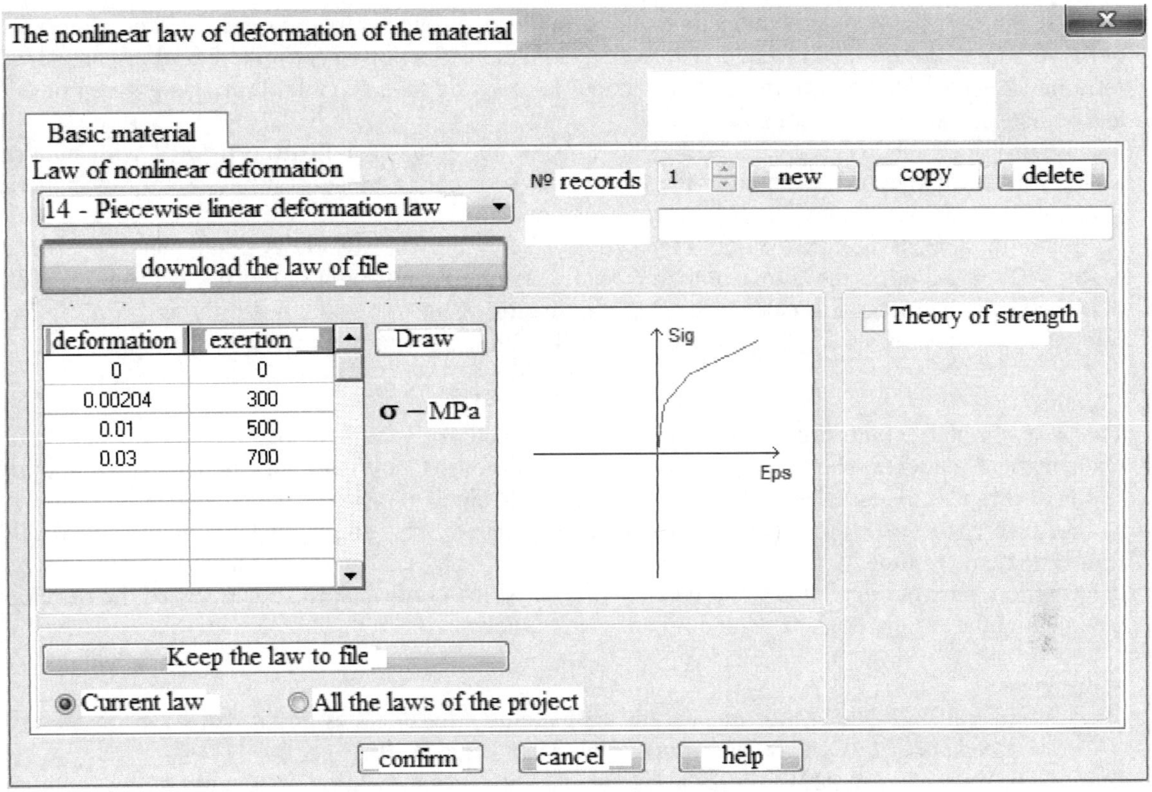

Figure 5. The law of nonlinear deformation of a material.

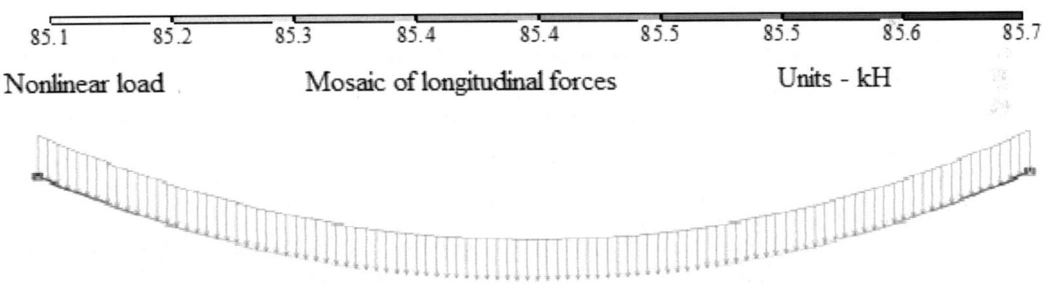

Figure 6. Mosaic of longitudinal forces.

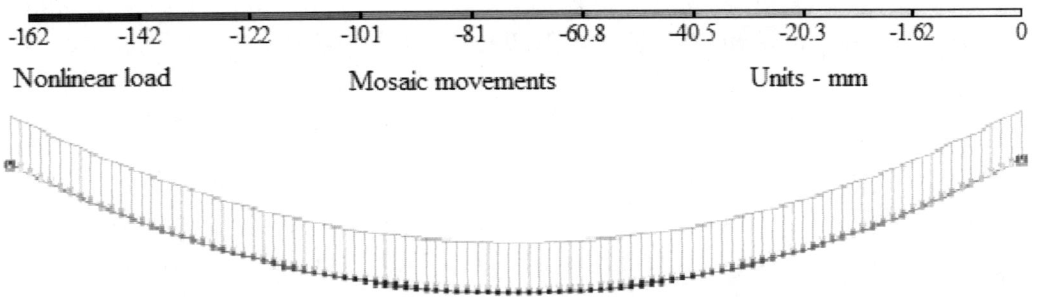

Figure 7. Mosaic movements.

IOP Conf. Series: Journal of Physics: Conf. Series **1084** (2018) 012008 doi:10.1088/1742-6596/1084/1/012008

As a result of mathematical modeling of SSS with the proposed method the corresponding values of the maximum longitudinal force $T_1(0) = T_1(l)$ and the deflection in the middle of the span $w(l/2)$ for a number of values of a uniformly distributed load acting on a flexible thread with given physical and geometric parameters were obtained.

The proposed method for calculating the stresses and strains of flexible threads with nonlinear elastic properties was implemented in the MathCAD 12 program by substituting the above numerical values of the parameters into the above formulas.

When implementing the method of calculating flexible threads of nonlinear elasticity in the MathCAD 12 program, the error compared to the finite element method did not exceed 5%. When implementing the method of calculating flexible threads of linear elasticity [9], the error in the Excel program did not exceed 6.5% in comparison with the finite element method.

Conclusion

The analysis of flexible thread SSS parameters obtained with the help of the proposed method of mathematical modeling showed quite a high degree of consistency with the calculation results in the LIRA-CAD software complex that implements the finite element method.

The discrepancies in the values of the parameters of the stress-strain state in comparison with the finite element method taken as the main comparison criteria for all sections of the accepted deformation diagram amounted to no more than 5%, which confirmed the adequacy of the developed method and the assumptions made in computer simulation.

References

[1] Velinsky S A 1993 A stress based methodology for the design of wire rope systems. *J. Mechanical Design, Transactions of the ASME* **115(1)** 69

[2] Cao X and Wu W 2018The establishment of a mechanics model of multi-strand wire rope subjected to bending load with finite element simulation and experimental verification. *Int. J. Mech. Sci.* **142** 289

[3] Xing E and Zhou C 2016 Analysis of the bending behavior of a cable structure under microgravity. *Int. J. Mech. Sci.* **114** 132

[4] Fontanari V, Benedetti M and Monelli B D 2015 Elasto-plastic behavior of a Warrington-Seale rope: Experimental analysis and finite element modeling. *Eng. Struct.* **82** 113.

[5] Jolicoeur C 1997 Comparative study of two semicontinuous models for wire strand analysis. *J. Eng. Mech.* **123(8)** 792

[6] ErdemImrak C and Erdönmez C 2010 On the problem of wire rope model generation with axial loading. *Math. Comput. Appl.* **15(2)** 259

[7] Šaraškinas V and Valiunas B 2007 Testing and calculating three-hinge suspension structures of hollow concrete-filled steel tubular members. *Technol. Econ. Dev. Eco.* **13(2)** 153

[8] Jolieoeur C 1996 A general formulation of the theory of wire ropes. *J. Appl. Mech-T. ASME* **63(4)** 1053

[9] Tarasov D A, Konovalov V V and Zaitsev V Y 2013 Mathematical modeling of the stress-strain state in steel ropes. *Vestnik Saratov State Technical University* **73(1)** 215, available at: https://elibrary.ru/download/elibrary_21579671_68169534.pdf

IOP Conf. Series: Journal of Physics: Conf. Series **1084** (2018) 012009 doi:10.1088/1742-6596/1084/1/012009

Mobile trailer support arm: modeling and analysis of breakage

V Zaitsev[1], V Konovalov[2], G Gumarov[3] and Y Rodionov[4]

[1]Department of Machine Building Technology, Penza State Technological University, Penza, 1A/11 proyezd Baydukova/Gagarina street, Penza 440039 Russian Federation
[2]Department of Machine Building Technology, Penza State Technological University, Penza, 1A/11 proyezd Baydukova/Gagarina street, Penza 440039 Russian Federation
[3]Institute of Additional Education, West Kazakhstan State University named after M. Utemisov, 162 Dostyk Avenue, Uralsk 090014 The Republic of Kazakhstan
[4]Department of Machine Parts, Tambov State Technical University, 106 Sovetskaya Street, Tambov 392000 Russian Federation

[1]E-mail: vluzai@gmail.com

Abstract. Calculation of vehicle elements load is a mandatory operation in the engineering of transport. The paper presents the calculation schemes of the static loading of the lever supports of the mobile hopper of bulk materials and determines internal force factors. The numerical values of equivalent stresses distributed on the lever of the mobile hopper of bulk materials are determined. The visualization of the results of calculating the stresses performed by the classical cross-section method and the finite element method is given. The results of both calculations show that the dangerous zone in the cross section of the lever is the neighbourhood of the hydraulic cylinder attachment. The analysis by the loading method shows the risk of breakage in the place of attachment of the hydraulic cylinder bracket from the side of the support wheel. To eliminate the revealed shortage there are proposals and recommendations given by the authors.

1. Introduction

Ensuring the required quality of product indicators is an urgent task of machine building and instrument making industry. Quality is a complex indicator, in the composition of which one can single out the working efficiency of machine-building product determined mainly by the main criteria as strength. Strength is laid at the design stage of the product. To do this the product is calculated for strength under static and dynamic loads at the checking calculation stage [1, 2]. An additional condition of operability is to ensure the rigidity of the structure [3, 4].

At present, strength calculation is widely provided by Computer Aided Design systems (CAD) [5, 6] assigning 3D models of the final product, external force factors (location of application, nature of the load, their numerical value or law of variation) and the boundary conditions. After setting necessary CAD system' parameters the specific calculations are performed by finite element methods giving visualization of a particular product with calculated strength values and estimated strength conditions [6].

However, despite the widespread use of strength calculations in CAD systems, classical methods for calculating strength and rigidity based on material resistance methods and elasticity theory may be more relevant [1-3].

Although numerical analysis [7] of the reasons causing the breakage of trailer support levers near the attachment of the hydraulic cylinder has been made, there is still a wide variety of structures of support arms used in machines and devices, and problems inducing mechanism failure at material exploitation.

The aim of the work was to reveal the most loaded sections of the support bracket of the mobile wheel trailer and the main possible places for its breakdown.

2. Theoretical research

We investigated the internal force factors arising in the support lever of the discrete materials bunker of the transport trailer. The loading and attachment conditions of the support lever under static conditions were justified in [7In this paper the variation of stress along the length of the support lever by theoretical methods was studied. To calculate the distribution of equivalent stresses along the axis of the lever we used the cross-section method. The working capacity of the support lever was tested according to the strength condition. The search for the zone of increased stress values of the lever was carried out under static loading. The maximum value of stresses did not exceed the allowed stresses taking into account the availability of the safety factor for multiple shock load. The calculation scheme for placing the support lever, for determining the external nature of loading, had the form shown in Figure 1.

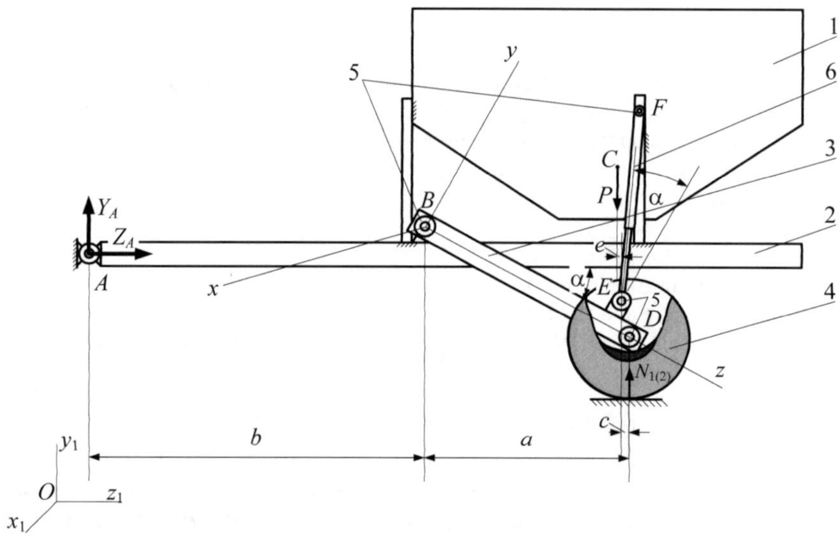

Figure 1. Scheme of the support lever static loading: 1 – bunker; 2 – frame; 3 – support arm; 4 – support wheel; 5 – basic hinge; 6 – hydraulic cylinder lifting the frame with a bunker; A – joint hinge of the trailer to the mover; B – joint hinge of the support lever to the frame of the trailer; C – the center of gravity of the frame with the trailer bunker; D – wheel axle; E – the joint hinge of the hydraulic cylinder to the support arm; F – the joint hinge of the hydraulic cylinder to the frame leg of the trailer.

For the presented design scheme, external force factors were determined that acted on the support lever under static loading with equation (1) determining the value of the load (N) required to define the reactions of lever supports for mobile bunker of loose materials:

$$N = \frac{P \cdot (a+b-e-c)}{2 \cdot (a+b)}, \tag{1}$$

where P – vertical projection of the force action of the trailer's weight to the point of attachment of the hydraulic cylinder to the support lever, *N*; a, b, c and e– distances between constructive elements (Figure 2), *m*.

The design diagram of the support arm 3 (Figure 1) as a figure of B, E, and D points lever in space, as shown in [7], can be represented in the form shown in Figure 2 used to obtain the dependencies between forces and moments.

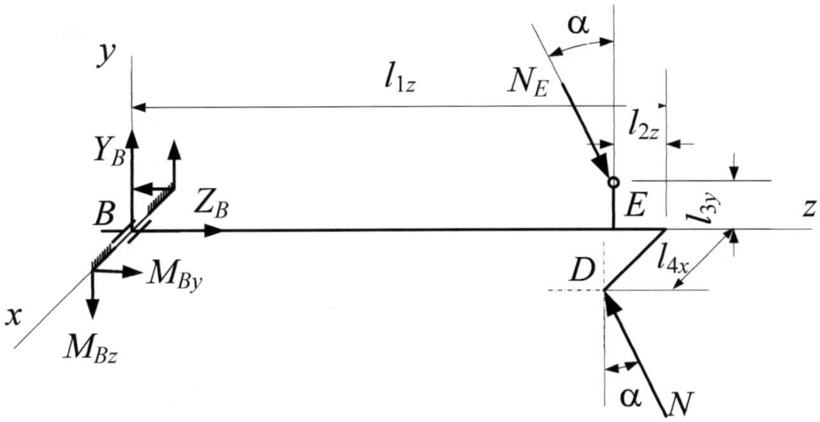

Figure 2. Calculation diagram of the support arm 3 of the mobile trailer (Figure 1): B – cylindrical hinge for mounting to the frame 2, E – spherical joint for hydraulic cylinder mounting, D – wheel support, N, NE, active force YB, ZB– reaction of the arm mount support, MBZ, MBY – moments bending along the Z and Y, α – angle between longitudinal axis of the cylinder and the normal to the lever, l1z, l2z, l3y, l4x – projection lengths of the lever on the respective axes.

According to Figure 2 the forces at the points of linking the lever with other elements of the structure (the reaction of external and internal bonds) shown in Figure 1 were determined. We calculated the force of hydrocylinder' rod action on a support lever as follows:

$$N_E = \frac{N \cdot \cos(\alpha) \cdot (l_{1z} + l_{2z})}{\cos(\alpha) \cdot (l_{1z}) + \sin(\alpha) \cdot (l_{3y})}, \tag{2}$$

moments bending along the Z and Y axes about the B axis were found as:

$$M_{Bz} = N \cdot \cos(\alpha) \cdot l_{4x} \text{ and } M_{By} = N \cdot \sin(\alpha) \cdot l_{4x}. \tag{3}$$

Further, we calculated the reaction of the supports along the axes at point B (Figure 1):

$$Y_B = N_E \cos(\alpha) - N\cos(\alpha) \text{ and } Z_B = -N_E \sin(\alpha) + N\sin(\alpha), \tag{4}$$

where α – angle between the longitudinal axis of the hydraulic cylinder and the normal to the lever, radian; N – balancing force of the wheel on the support arm, N; l_{1z}, l_{2z}, l_{3z}, l_{4x} were the lengths of the structural element (Figure 2).

We used the obtained values of external force factors to determine internal force factors (longitudinal forces, transverse forces of the corresponding projections of the lever sections, bending

IOP Conf. Series: Journal of Physics: Conf. Series **1084** (2018) 012009 doi:10.1088/1742-6596/1084/1/012009

moment, and torque in the corresponding hinges) arising at the corresponding points of the lever sections.

The longitudinal forces along the z axis for point D and E are defined as:

$$N_{zD} = -N \cdot \sin(\alpha); \quad (5)$$

$$N_{zE} = -N \cdot \sin(\alpha) + N_E \cdot \sin(\alpha); \quad (6)$$

The transverse forces along the x and y axes are defined as:

$$Q_{xD} = N \cdot \cos(\alpha); \quad (7)$$

$$Q_{yD} = -N \cdot \cos(\alpha); \quad (8)$$

$$Q_{yE} = -N \cdot \cos(\alpha) + N_E \cdot \cos(\alpha). \quad (9)$$

The bending moments along the x and y axes are defined as:

$$M_{yD} = M_{By}; \quad (10)$$

$$M_{xB} = 0, \quad (11)$$

$$M_{xE} = Y_B \cdot l_{1z}, \quad (12)$$

$$M_{xD} = Y_B \cdot \left(l_{1z} + l_{2z}\right) - N_E \cdot \cos(\alpha) \cdot l_{2z} + N_E \cdot \sin(\alpha) \cdot l_{3y}; \quad (13)$$

where Z axis torque is defined as:

$$M_z = M_{Bz}. \quad (14)$$

Analysis of the design scheme and expression for internal force factors allowed us to conclude that the lever had a complex form of loading - tension and twist with torsion.

In this case, the strength condition of the product was determined by the third or fourth strength theories for the principal stresses:

$$\sigma_\Sigma = \frac{N}{A} + \frac{M_\Sigma}{W} \le [\sigma], \quad (15)$$

where σ_Σ – equivalent normal stress, N – longitudinal force, A – cross-sectional area of the lever, M_Σ – equivalent moment, $[\sigma]$ – allowable stress; W – the axial moment of the section resistance for the profile pipe with respect to the x or y axis, which were respectively determined as follows:

$$W_y = \frac{h \cdot b^2}{6} - \frac{h_1 \cdot b_1^2}{6}, \text{ and } W_x = \frac{b \cdot h^2}{6} - \frac{b_1 \cdot h_1^2}{6}, \quad (16)$$

where h the height of the outer contour of the pipe, b the width of the outer contour of the pipe, h_1 the height of the inner contour of the pipe, b_1 the internal pipe width; $A = b \cdot h - b_1 \cdot h_1$ cross-sectional area of the profile pipe.

For the "third" and "fourth" theories of structural strength equivalent moments were calculated by means of equations (17) and (18), respectively:

$$M_\Sigma = \sqrt{M_\sigma^2 + M_\tau^2}, \quad (17)$$

$$M_\Sigma = \sqrt{M_\sigma^2 + 0.75 \cdot M_\tau^2}, \quad (18)$$

where M_σ, M_τ – bending and twisting moments.

3. Results and discussion

The analysis of equivalent stress diagram (Figure 3) showed that the largest value of equivalent stresses (about $1.304 \times \sigma \times 10^8$ Pa) was located near the point E (Figure 2). Therefore, the most

IOP Conf. Series: Journal of Physics: Conf. Series **1084** (2018) 012009 doi:10.1088/1742-6596/1084/1/012009

dangerous cross-section of the lever was the cross section in the neighbourhood of the hydraulic cylinder mounting. As the coordinates of the lever approached the support point B (Figure 2), the stresses slightly decreased. When the lever coordinates approached the support point D (Figure 2), the magnitude of the stresses dropped sharply and substantially.

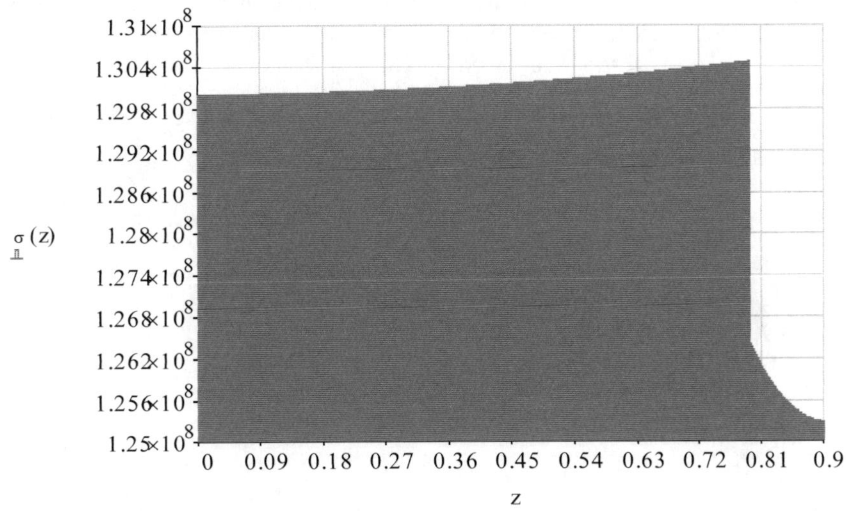

Figure 3. Diagram of the distribution of equivalent stresses along the axis of the lever.

Equivalent stress distribution diagram σ (Sigma, PA) of lever 2 (Figure 2) along its axis for points at a distance z=[0, l_{1z}+l_{2z}] (m) from point B. The obtained results were confirmed by the calculation of the lever by finite element method (Figure 4): the hinge mount of the lever to the frame (point B, Figure 1) was the main support of its attachment. With respect to the joint in the given point, the lever was deformed. Considering the hinge mount in points B and E (Figure 1), the fit in these hinges was made with a gap. In addition to the nominal design gap size, there was an actual clearance due to tolerance size and the normal size distribution law in the process of manufacturing the parts.

Thus, the presence of gaps in the conjugation leads to the possibility of a lever rotation in point B around z axis. As a result of combination of possible turns of the lever elements at these B and D points, two ways are possible as to include or exclude a reserve of possible rotation of the hinge around the z-axis in point E. The results of modeling for both variants of joining the lever using the finite element method performed in the CAD system are presented bellow.

Figure 4 shows the results of calculating the stresses distributed over the lever in accordance with the indicated options. When moving from B to E point (Figure 4a), the color turned from blue into green that confirmed the stresses growth and the greatest stresses was near the upper brace (for E point) of the cylindrical support in point D. There appeared a small area of the surface with a shade of yellowness (Figure 4a). The transition of color from green to yellow or further to red, indicates an increase in The results obtained corresponded to analytical calculations. The largest stresses are located near point E in Figure 2.

In the second version of the calculations, when moving away from point B, the stress value also went up(color changed from blue to green on Figure 4b), but in E point it faded. The cylindrical hinge in point E assumed a load from the twisting of the lever. Near point E, at the upper brace there was also a platform on the side face with increased stresses (green color). The specified place of increased stresses for the two loading options worked for scrapping due to the oscillations in the lever during its operation.

IOP Conf. Series: Journal of Physics: Conf. Series **1084** (2018) 012009 doi:10.1088/1742-6596/1084/1/012009

In addition, the twisting of the lever resulted in the transfer of the load from the pair of brackets of the hinge (point E) to the near bracket, that further caused the stress in the stress concentration zone (Figure 1).

The bend of the hinge axis in point E could also affect the duration of the normal operation of the hydraulic cylinder due to the bending of the rod. When the trailer moved depending on the magnitude of the applied load (the presence of fluctuations in the magnitude of the applied force in point D), the alternating operation of the lever in both loading modes was possible (Figure 4).

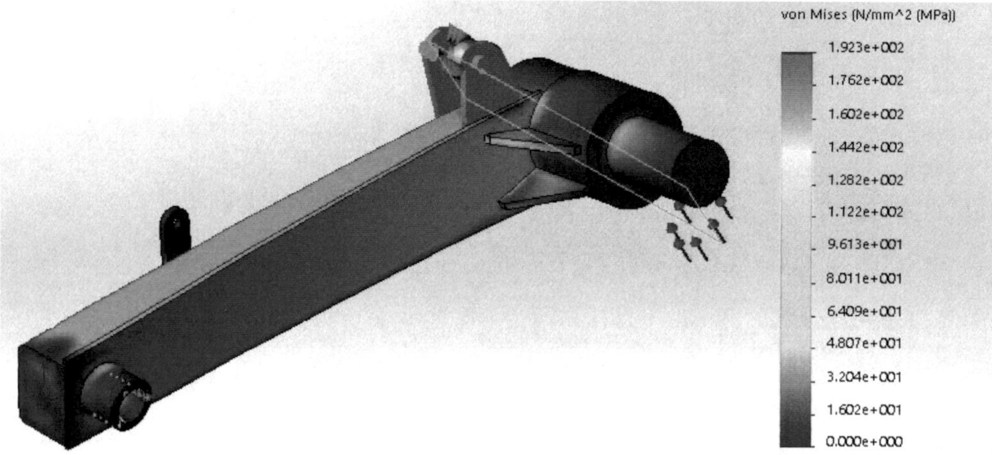

a)

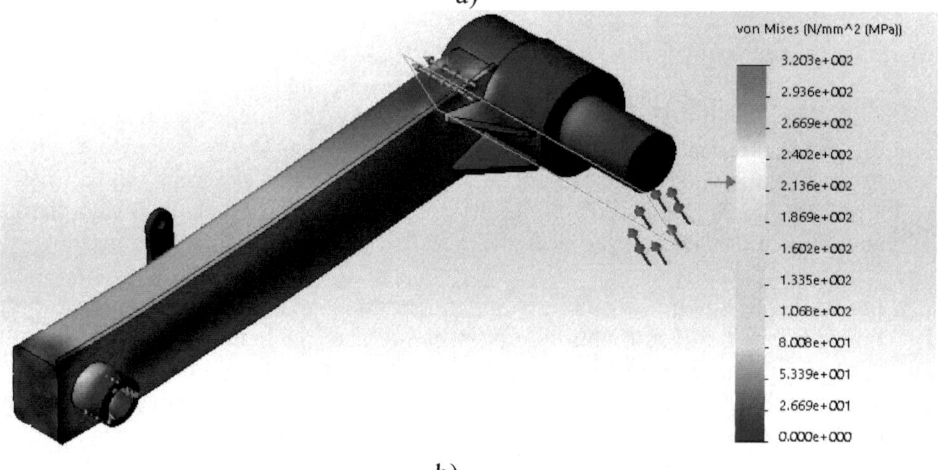

b)

Figure 4. Distribution of stresses along the arm for attaching (a) and joining (b) the lever to point B on cylindrical faces with force applied at an angle α at point D and connector at point E of spherical bearing.

Conclusion

In the paper the calculation of the support lever of a transport trailer was made. Then the analysis of possible impacts of supporting wheels on the supporting arm made it possible to establish the causes and the place of support levers breakdown during vehicles exploitation, and, further, to give to recommendations for breakdowns elimination.

It has been shown that regardless of calculation method, there is a dangerous zone in the cross section of the lever in the vicinity of the joint of the hydraulic cylinder. Analysis of loading methods and unavoidable load fluctuations in the high stress zone has demonstrated the danger of breakage in

IOP Conf. Series: Journal of Physics: Conf. Series **1084** (2018) 012009 doi:10.1088/1742-6596/1084/1/012009

the place of hydraulic cylinder bracket joint from the side of the support wheel, which was confirmed by breakdowns in some vehicles.

To eliminate the revealed shortage, it is necessary to avoid the appearance of bending moments (analogs to M_{Bz} and M_{By}) in the axis of fastening of the hydraulic cylinder. To do this, we propose to use the spherical support of the hydraulic cylinder or increase the clearance in the hinge at the outer bracket. Alternative but not the best way may include the increase of the lever rigidity.

References

[1] Alyushin Yu A 2003 Specific features of power analysis of the rocker mechanisms *Problemy Mashinostraeniyai Nadezhnos'ti Mashin* **3** 9

[1] Kutlubaev I M, Zhydenko I G and Bogdanov A A 2016 Basic concepts of power anthropomorphic grippers construction and calculation. *2nd Int. Conf. Industrial Engineering, Applications and Manufacturing, ICIEAM 2016* **7** 910

[2] Ling M, Cao J, Jiang Z, Zeng M and Li Q 2018 Optimal design of a piezo-actuated 2-DOF millimeter-range monolithic flexure mechanism with a pseudo-static model *Mechanical Systems and Signal Processing* **115** 120

[3] Gong J, Pang J and Zhang Y 2014 Stiffness analysis of a compliant mechanism considering complex deformations of all flexible structure units *Micro and Nanosystems* **6(3)** 156

[4] Lai L J and Zhu Z N 2017 Design, modeling and testing of a novel flexure-based displacement amplification mechanism *Sensors Actuat. A* **266** 122

[5] Jomartov A 2017 Multibody dynamic model of Russian sulzer loom STB *J. Vibrational Eng.Technol.* **5(6)** 523

[6] Vavro J, Vavro J, Pecušová B and Burget M2018 Dynamic analysis of lever mechanism for manufacturing of raw tyres *Manufacturing Technol.* **18(1)** 145

[7] Zaitsev V, Konovalov V, Vol'nikov M and Petrov A 2017 Reactions determination of the wheel trailer mounting linkage *Samara State Agricultural Academy Bulletin* [in Russian] **4** 50, available at: http://ssaa.ru/structur/riz/izvestia/vypuski/2017_4_ru.pdf

IOP Conf. Series: Journal of Physics: Conf. Series **1084** (2018) 012010 doi:10.1088/1742-6596/1084/1/012010

Optimization in work modeling of a mixer

A Chupshev[1],V Konovalov[2] and M Fomina[1]

[1]Faculty of Engineering, Penza State Agrarian University, 440014 Penza, Botanical street d. 30 Russian Federation,
[2]Faculty of Industrial Technologies, Penza State Technological University, 440039 Penza, travel Baydukova/Gagarin street, 1A/11 Russian Federation

[1]E-mail: chupshev.a.v@pgau.ru

Abstract. The paper presents a list of indicators used to optimize the working process of rotary machines using the example of a mixer. The shortcomings of polynomial statistical models are indicated and the use of functions which nature of variation corresponds to the theoretical positions is recommended. The theoretical dependencies of the mass of the prepared portion of the mixture, the quality of the mixture, the duration of mixing, the productivity of the mixer, the drive power, the specific energy consumption are revealed, taking into account the uniformity of the mixture. These dependencies and the refinement coefficients that appear in them have the form of power and exponential functions. These functions do not have an extremum, therefore optimal parameters are impossible.

1. Introduction

Devices providing formation of various mixtures and composite substances are widely used in industry, construction, and agriculture. There are numerous studies justifying the conditions for a fast and efficient process of mixing components at low energy costs [1-4]. Rotation of working bodies acting on different material media are generally placed inside a container of the investigated machines [1, 2]. To describe working process of substances moving and mixing, specialist reveal many dependences as theoretical (based on the laws of mechanics and hydraulics [5]) and/or functional (statistical) ones based on the justification of quantitative (productivity/flow, duration of the mixing cycle), qualitative (uniformity / uniformity of the mixture, average particle size, temperature, etc.) and energy (drive power, specific energy consumption / energy consumption of the process) indicators of the process flow from the dimensional characteristics of the working bodies [1-5]. Agricultural mixers, according to the requirement of regulatory and normative documents in Russia and Belarus, have to follow controlled parameters like productivity, process duration, power and energy consumption, and unevenness, the coefficient of variation of the distribution of the control component in 15-20 samples of a given mass and concentration of the check component. By analogy with other devices [1-5], the operation of mixers was simulated, and indicators were monitored on the basis of statistical dependences [6].

According to the theory of experimental multifactor planning [6], processing the results of interpolation studies needs polynomial dependences which describe technical processes [6, 7]. The linear model is firstly tested when describing the data. If the obtained results are not adequately described by linear model, it is necessary to proceed to the regression equations of the second and subsequent orders. With an adequate description of the results revealed by the model, a researcher

IOP Conf. Series: Journal of Physics: Conf. Series **1084** (2018) 012010 doi:10.1088/1742-6596/1084/1/012010

should remove insignificant coefficients in the polynomial regression equation from the obtained order equation using the Student's test, thereby simplifying, if it is possible, the given equation.

Obtained functional statistical expressions do not reflect the physical essence of the phenomenon, but due to the refinement of the coefficient values, the regression equations tend to compress the reference surface of the model to experimental values with a minimum error. Extrapolation of the revealed dependence beyond the experimental limits is practically impossible [1].

It is known [8-10] that the power expended on the work process adequately describes the power functions, and exponential functions are applicable to describe the change in the quality of the mixture. These dependencies are of a functional nature, not reflecting the physical essence of the process studied. Therefore, to understand the physical essence of the phenomenon, one should use theoretical expressions.

The aim of our work is to establish the type of dependencies theoretically describing the working process of mixing components in a mixer and to reveal effective parameters of the mixer construction design.

2. Experimental Part

The quality index of the mixture, the cost of power, the energy consumption of mixing, and the productivity in the mixing cycle were studied in the research.

The research methodology provided for a number of actions. In order to achieve the aim we firstly established main indicators of the working process of the stirrer on the basis of theoretical studies analysis [9, 10]. Then we found the extremum (optimum) for the indicators of the stirrer working process. We analyzed the data to determine the effective performance of the stirrer process as a criterion for process optimization. Finally, we analyzed the possibility of description of technical and technological processes by polynomial expressions.

Based on the known information [9-11], we describe the working process of the agitator-mixer (as an example of rotary devices) on the basis of theoretical expressions with the introduction of additional correctional coefficients and we will try to optimize the parameters of the technological process.

The mass of the portion of the prepared mixture M is a particular case of the power function, where the exponent is equal to "one" and is described, kg:

$$M = Vo \cdot E \cdot \rho. \qquad (1)$$

where V_o– is the volume of the mixing tank, m^3; E – is the degree of filling the mixing tank by the ingredients, shares; ρ – is the density of the heap of the mixture, kg / m^3.

A number of authors propose to describe the mixing process by analogy with the diffusion process by an exponential function. Uniformity of the mixture (0.01%):

$$\theta = 1 - e^{-k \cdot Tc}, \qquad (2)$$

where k – is the empirical coefficient of mixing intensity; Tc – is the mixing time of the material, s.

The uniformity of the mixture (0.01%) (as "relative uniformity" [12]) can be written with the help of the coefficient of variation ν (0.01%) of the content of the control component in the samples:

$$\theta = 1 - \nu, \text{ or } \nu = e^{-k \cdot Tc}, \qquad (3)$$

For different mixers, the empirical mixing intensity k will be different [9, 10]. Therefore, the graphs of variance of the uniformity of the mixture Q for such mixers will differ in intensity (Figure 1). The uniformity of the mixture before mixing is $Q_н$, required at the end of mixing – $Q_к$. The current values of the uniformity of the mixture for different mixers are Q_1 (T) and $Q_2(T)$.

To minimize energy costs and financial costs, it is desirable to shorten the mixing time ($[T_к$ - $T_н]$ and $[t_к$ - $t_н]$). Accordingly, if we know the values of k of several mixers or their operating regimes, we can find the required mixing time for ingredients in a particular case.

The conditional time before the mixing is determined by:

$$T_н = -\frac{1}{k} \cdot \ln(1-\theta_н), \tag{4}$$

The conditional time before the end of mixing is determined by:

$$T_K = -\frac{1}{k} \cdot \ln(1-\theta_K), \tag{5}$$

The required mixing time of the components in the mixer in general form is determined as follows:

$$T_C = T_K - T_H = -\frac{1}{k} \cdot \ln\frac{(1-\theta_K)}{(1-\theta_H)}. \tag{6}$$

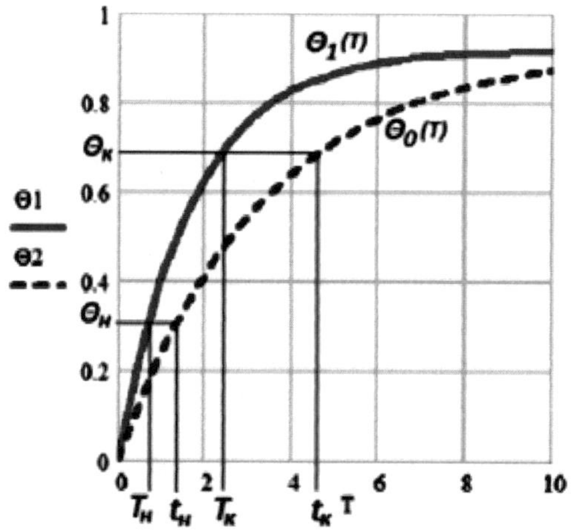

Figure 1. Change in the uniformity of the mixture Q (0.01%), depending on the mixing time T, s.

According to the technological requirements for the quality of the mixture, a number of uniformity values of the mixture are regulated as critical (permissible) values: 80%, 90%, 95%. Then: $\Theta_K \geq \Theta_{zoo}$, where Θ_{zoo}=0,80≅ v_{zoo}=0,20; Θ_{zoo}=0,90≅ v_{zoo}=0,10; Θ_{zoo}=0,95≅ v_{zoo}=0,05. In this case, the mixing time should be, s:

$$T_C \geq \frac{-\ln(v_{zoo})}{k} = \frac{1}{Sk_i} \tag{7}$$

where Sk_i –is the total factor of the empirical coefficients of the mixing duration, taking into account the constructive, kinematic and technological features of the work, given, for example, in the work [10] in the form of power functions.

The form of this expression corresponds to a "hyperbola" which is the particular case of a power function with a negative value. The productivity of the mixing tact of the ingredients in the mixer is a special case of the power function [10], kg / s:

$$Q_C = \frac{M}{T_C} = \frac{V_o \cdot E \cdot \rho}{T_C}. \tag{8}$$

In mechanics the power is determined by:

$$N = F \cdot \upsilon, \tag{9}$$

where F – is the motion resistance force, H; υ –is the displacement velocity of the object in the environment, m/s.

In the resistance of materials, the voltage is determined by [6], Pa:

$$\sigma = F/S\text{, where the force } F = \sigma \cdot S, \quad (10)$$

where σ – working stresses, PaN/m^2; S – is the area of force application, m^2.

The area of application of force is determined by, m^2:

$$S=(l_s \cdot \sin(\alpha)) \cdot dR \cdot Z=(K_1 \cdot D \cdot \sin(\alpha) \cdot Z) \cdot dD, \quad (11)$$

where l_s – is the blade width, m; a – the angle of attack when installing the blade, rad .; dR – the length of the current radius section of the force application, m; Z – the number of blades, pcs; D – the diameter of the mixing tank, m; K_l – the coefficient of proportionality of the width of the blade and the diameter of the tank; dD – the length of the section of the current radius of the force application, expressed in terms of the diameter, m.

The velocity of the end of the blades during rotation is described by, m/s:

$$\vartheta = \omega \cdot R = \omega \cdot D/2, \quad (12)$$

where ω – is the angular velocity of the rotor, rad/s; R – is the maximum radius of rotation of the point of application of force, m.

Considering the fact that the radius of the force application for the blades of mixer is not constant, the power equation for mixing the components in the mixer will be written as follows:

$$N = \int_0^{D/2} \{[\sigma \cdot (K_l \cdot D \cdot \sin(\alpha)) \cdot dD \cdot Z] \cdot (\omega \cdot D/2)\} dD = K_l \cdot \sigma \cdot Z \cdot D^3 \cdot \sin(\alpha) \cdot \omega/(2 \cdot 3) \Big|_0^{D/2}. \quad (13)$$

Power at mixing is determined as follows, W:

$$N = \pi \cdot K_l \cdot \sigma \cdot Z \cdot \left(\frac{D}{2}\right)^3 \cdot \sin(\alpha) \cdot n/180, \quad (14)$$

where n – is the rate of rotation of mixer, min^{-1}.

The power expression is a particular case of a power function, where the exponent of some of the indicators exponents is equal to "one" (Figure 3). The value of the stress O is very difficult to determine theoretically, therefore it can be found experimentally as a correction factor in the power expression 14. For example, for a blade mixer where $K_l = D / 6$, the exponent is determined by the power function:

$$\sigma = 10125{,}86 \cdot M \cdot D^{-4} \cdot n^{-0{,}185863} \cdot Z^{-0{,}606518} \cdot \sin(\alpha)^{0{,}653631} \cdot (ls^{0.543368} + L^{0.962341}) \quad (15)$$

where L – is the length of the blade, m; ls – is the width of the lining plate on the blade, m.

Specific energy consumption per tact of mixing is a power function of the indicators, J/kg:

$$Y = \frac{N \cdot Tc}{M} = \frac{\pi \cdot K_l \cdot Z \cdot \sin(\alpha) \cdot n \cdot (D/2)^3 \cdot Tc}{180 \cdot V_o \cdot E \cdot \rho}, \quad (16)$$

It is also possible to define the optimization criterion as the specific energy consumption per tact, taking into account the uniformity of the mixture, J/kg:

$$Y_k = Y/(\theta) = Y/(1-\nu). \quad (17)$$

The specific energy consumption of mixing, taking into account the uniformity of the mixture for the mixer at the current value of uniformity Θ is determined by:

$$Y = \frac{N \cdot Tc}{\theta \cdot M} = \frac{\pi \cdot K_l \cdot Z \cdot \sin(\alpha) \cdot n \cdot (D/2)^3 \cdot Tc}{180 \cdot \theta \cdot V_o \cdot E \cdot \rho}, \quad (18)$$

Analyzing the formulas (1...14), it is clear that they are based on a number of indicators: constructive (V_o, D, Z, K_l, α), kinematic (n) and technological (E, Tc, ρ). Some indicators (σ, k)

depend both on the physico-mechanical properties of the material of the mixture and on the design-regime parameters, i.e. are an empirical power function of these arguments.

We will analyze the function of the energy intensity of mixing, taking into account the uniformity of the mixture (14) as the most complex one which comprised of complex of other specified functions of the mixing process. This function, as well as other functions, is a function of several variables which were specified earlier.

To confirm the optimal values of design-regime and technological parameters, these functions must be checked for externality.

If the differentiable function $y = f\ (x_1, x_2, \ldots_n)$ has an extremum $tM_o = (x_{10}, x_{20}, \ldots\ x_{no})$, then all its parts of the first-order derivative at this point are equal to zero:

$$\frac{\partial y}{\partial x_1}\bigg|M_0 = \frac{\partial y}{\partial x_2}\bigg|M_0 = \cdots = \frac{\partial y}{\partial x_{n-1}}\bigg|M_0 = \frac{\partial y}{\partial x_n}\bigg|M_0 = 0, \tag{19}$$

By the necessary condition of extremum of function of several variables, all its partial derivatives at the extremum point must be equal to zero.

3.Results and discussion

These functions (1...14) are the product of several functions of one variable. Therefore, all its partial derivatives (functions of one variable) are exponential and monotonically increasing or decreasing. Consequently, the functions (14) do not have an extremum at positive values of the arguments (*Vo, D, Z,* K_l≅*ls, α, n, E, Tc, ρ, σ*).

The optimization of the process is possible if the mixer parameters are experimentally justified, when the stable steady pattern of motion is disturbed (the growth of a certain parameter of the blade / mixer / starts to interfere with the movement of the material, which makes the quality of the mixture worse, i.e. it changes the nature, proceeding and movement of material in the mixer tank). These cases are determined by the features of the functions of the coefficients Sk_i and σ.

For example,other constant values of the parameters, increasing the height of the blade of the vertical mixer with the minimum blade height does not set in motion the entire mass of material in the tank. As the height of the blade increases, all material begins to move and mix. Further growth of the blade height will lead to the formation of sectors that prevent typical free movement of the material, the material will accumulate inside with minimal mixing, which will increase the duration of mixing and the energy intensity of the process. In this case, we can see the possibility of an extremum with increasing blade height. However, such an experiment is counterproductive. Another algorithm of actions is more effective.

The blade height can be different with a nominally loaded engine of a constant installed power, but the lowering of the altitude will increase the speed of the mixer, which will increase turbulence, reduce processing time, increase productivity and, possibly, reduce energy intensity. We will get a set of combinations of parameters where it will be possible to find the highest device performance or the lowest energy costs and thereby rational values of constructive, kinematic and technological parameters by means of exhaustive search method with the loaded engine.

To describe the working processes and to find the optimum, it is widely recommended to use polynomial functions [6]. However, there are questions to the validity of the application of these expressions.

The resulting polynomial models of the process will seek to obtain an extremum, which the functions will seek to obtain forcibly. In this case, the optimization problem is to find the min / max values. At the same time, the nature of the experimental values may not have an extremum.

An example of the absence of extremum is a number of processes previously studied and described [9, 11, 12]. As the number of impacts increases, the quality of the mixture will improve. At the same time, there is no possibility of effective separation. The coefficient of variation of the content of the component in the sample will decrease. The values of the coefficient tend to zero, slowing to zero without reaching it as shown in [11, 12]. When cleaning the surface, the mass of dirt on the surface to be cleaned will decrease with the increase in the number of treatment effects. New pollution to take at the same time nowhere. It turns out that the second, increasing branch of the parabola cannot grow in principle on the graph of the experiment as demonstrated previously [9]. As the column of material in the vertical container increases, the pressure on the bottom will increase. At the same time due to friction of the material on the walls of the tank, the pressure increase will not be linear. The nature of the change in pressure corresponds to the exponential function [9].

The inexpediency of the application of polynomial equations is well seen from [9], the data of which is in Figure 2. To process the results of the pressure data (using its analogue – the force of the column of material P on the bottom of the tank) at a low height of the column of material H (points on the graph near zero, up to 50 cm) different formulas can be used. In particular, linear dependence, a second-degree polynomial, a third-degree polynomial, or exponential function (Table 1).

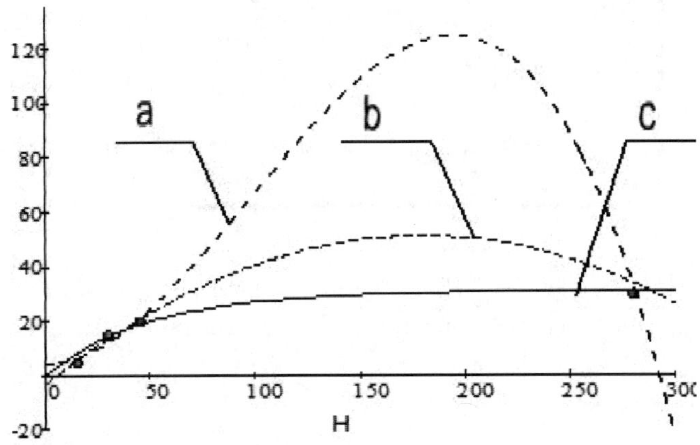

Figure 2. Influence of the height column (H, cm) on the pressure (P, Pa) by the third-order polynomial (a), the second order (b), and the exponential function (c).

To control the values, an additional point at 270 cm was taken. After processing the data by Statistica 5.5, the expressions in Table 1 were obtained. The results are presented in Figure 2. Height of polynomial graphs (Figure 2a and 2b) initially increases at different rates, and then falls down to negative values. In reality, it is impossible for the values of the material column pressure to the bottom.

Table 1. Influence of the regression equation on column pressure of feed.

Material	Type of model	Index of correlation with experimental data, %
Stockfeed	Linear function P=10.9+0.071· H	72.96
	Second-degree polynomial $P = 1.33 + 0.485 \cdot H - 0.0015 \cdot H^2$	98.40
	Third-degree polynomial $P = -1.89 + 0.73 \cdot H - 0.005 \cdot H^2 - 0.0001 - H^3$	99.86
	Exponential function $P = 35.78 \cdot (1 - e^{-0,02 \cdot H})$	99.13

The character of the exponential function increases (Figure 2c). The increase in values gradually decreases. The character of the change in the exponential function is close to theoretical dependences on the basis of the laws of theoretical mechanics. Such regressional exponential function can predict values in neighboring plots. In this case there occurs an error. Thus, problems of extrapolating functions to neighboring plots with the experimental zone are visible, despite the high statistical coefficients, e.g. correlation coefficient in Table 1. The highest correlation coefficient was for the third-degree polynomial. The exponential function also had a high value. The pressure of the material column height was also theoretically described by the exponential function. Therefore, the exponential function should be preferred.

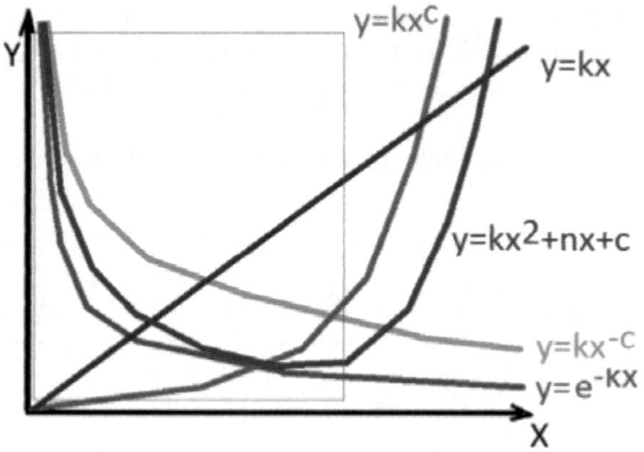

Figure 3. Change of Y from the value of factor X for functional dependencies of different types (genus): $y = e^{-k.x}$- of an exponent; $y = k - x^c$ - a power function (positive power); $y = k - x$ - a linear function (the power of the power function is equal to one); $y = k - x^{-c}$ - a hyperbola (the power of the power function is equal to the negative value); $y = k - x^2 + n - x + c$ - a parabola.

Thus, possible restrictions on the use of polynomial expressions were established. An analysis of the nature of the change in the arguments of the functions (Figure 3) for various functions showed that for small interval of study of indicators (red square), experimental points can be described by equations both power and polynomials. With the expansion of the research interval, there are contradictions between functions. Therefore, an incorrectly established relationship (the form of which does not correspond to theoretical expressions) can lead to incorrect results of the mixer studies. In this case, the expressions used for the statistical description of the process should have the form close to the theoretical dependencies.

Conclusion

Functional expressions describing the working process of the mixture formation have been established. The impossibility of theoretical determination of the optimum of the stirrer parameters was revealed. Functional expressions describing the mixer's working process are power or exponential functions that in principle do not have an extremum.

A method for finding rational valid values of the mixer parameters on the basis of the engine's load, observing the standards for the quality of the mixture, and finding the highest productivity or lowest power consumption of mixing from a possible combination of the values of the mixer parameters was proposed.

The impossibility of finding the extremum of the expressions for the mixer's working process and some series of processes due to the use of polynomial expressions that experimentally statistically describe these processes was revealed. The expressions used for the statistical description of the process should have the form close to the theoretical dependences.

References

[1] Pezo M, Pezo L, Jovanović AP, Lončar B and Kojić P 2018 Discrete element model of particle transport and premixing action in modified screw conveyors. *Powder Technol.* **336** 255

[2] Pezo L,Pezo M,Jovanović A, Banjac V,Đuragić2018O The joint mixing action of the static pre-mixer and the rotating drum mixer – Discrete element method approach *Adv. Powder. Technol.* **29(7)** 1734
[3] Leś K, Kowalski K and Opaliński I 2015 Optimisation of process parameters in high energy mixing as a method of cohesive powder flow ability improvement *Inz. Chem. Procesow.* **36(4)** 449
[4] Li W, Guo H, Huang Q, Hou Y and Zou W 2018 Effect of stirring rate on microstructure and properties of microporous mullite ceramics *J. Mater. Process. Techn.* **261** 159
[5] Tamrin K F, Sheikh N A and Rahmatullah B 2016 Numerical analysis of swirl intensity in turbulent swirling pipe flows *Sci. Eng.* **78(5-10)** 133
[6] Bilodeau M and Brenner D 1999 *Theory of Multivariate Statistics (*New-York: Springer-Verlag) p 289
[7] Jiang W 2011 *Asymptotic Theory of General Multivariate GARCH*, PhD Thesis, University of Western Ontario.
[8] Chemezov D, Tyurina S, Bayakina A and Lukyanova T 2018 Multi-factor experiment to determining of vibrations of steel pipes induced by vortex of air flow. *ISJ Theoretical & Applied Science* **03(59)** 201[in Russian]
[9] Konovalov VV 2003 *Workshop on the processing of the results of scientific research with the help of PC.* (Penza: RIO PGSHA)176 [in Russian]
[10] Ivanets V N, Bakin A I and Ratnikov S A 2004 *Processes and devices of food production: a Training manual* (Kemerovo: Kemerovo Institute of Food Technology) [in Russian] p 180, available at: http://dump.vstu.ru/files/storage/Kafiedry/MAPT/Biblioteka_uchebno_metodicheskoy_literatury/PAPP/Metodicheskaya_literatura/processy_ivanec.pdf
[11] Fomina M, Konovalov V, Cupshev A and Teruskov V 2016 Modeling of capacity vertical paddle mixer on the basis of statistical expressions [in Russian] *Innovative Techniques and Technol.* **3/08** 50
[12] Konovalov V, Cupshev A and Fomina M 2016 Modeling of the quality change of the blade mixer mixture on the basis of technological parameters *Innovative Techniques and Techno.* [in Russian] **3/08** 56

IOP Conf. Series: Journal of Physics: Conf. Series **1084** (2018) 012011 doi:10.1088/1742-6596/1084/1/012011

Analysis method of the information value of indicators

Y Artamonov[1] and I Kamanin[2]

[1]Federal State budget scientific institution "State scientific-methodological center", 51 Lyusinovskaya Street, Moscow 115998, Russian Federation
[2] Department of Economics of urban and housing rights, Moscow Metropolitan Governance University, 28 Sretenka Street, Moscow 109045, Russian Federation

[1]E-mail: junaart@mail.ru

Abstract. A new method for assessing the information significance of a group of indicators is proposed based on minimizing the matrix norm of deviations of the evaluation matrix from the matrix of the most informative indicators. The proposed method allows to combine the classes of tasks of autoinformativity and external informative characteristics. Comparison of the method with existing methods makes it possible to assert that in the proposed approach the information loss of the initial evaluation matrix is minimized. The analysis of regression dependencies on various subsets of indicators made it possible to conclude that in the developed method the construction of the most accurate linear regression dependencies is possible on the selected subset of informative indicators.

1. Introduction

A wide range of management tasks in education requires the selection of the most informative indicators. At the present day, the indicative method is widely used for managerial decision-making: the group of the most important in some sense indicators is selected, on which threshold values are being set. Then the objects of analysis are evaluated in a variety of ways according to the values of the selected indicators. The assessment of performance of universities, scientific organizations is conducted this way. A similar methodology is used to evaluate scientific and technical projects under various state programs in the field of science and education.

The task of determining the most informative indicators is similar to the problem of dimension reduction. However, in this case, it is required to select from the wide range of indicators the most important indicators, and not to suggest various transformations of the initial indicators, for example, in the form of linear convolution (as in the principal component method). In general, the scientific approaches used of dimension reduction can be divided into two classes:

class of methods based on the autoinformativity criterion aimed at maximum preservation of the information contained in the original data array for which each indicator is estimated by the level of influence on the overall variability of the data;

class of methods based on external information, aimed at maximum "extrusion" the information contained in the source array with respect to some other (external) indicators.

The first class of methods is based on various matrix decompositions. The key method in this case is principal component analysis (PCA) and its improvements [1-3]. An important and one of the first results in this direction was the theorem of C. Eckart, G. Young [4], in which it is shown that for a fixed rank of an approximation matrix the minimum of the Frobenius norm ensures the corresponding singular-value decomposition of the original matrix. Recently, methods of approximation of the initial

IOP Conf. Series: Journal of Physics: Conf. Series **1084** (2018) 012011 doi:10.1088/1742-6596/1084/1/012011

evaluation matrix by matrices of smaller ranks have become increasingly popular: non-negative matrix factorization (NMF), and various tensor decompositions options as well [5-8]. However, it is possible to use these methods for the task of selecting the most informative indicators only after an appropriate interpretation, which is often ambiguous. For example, for PCA as measure of the importance of the indicator factor loading of corresponding component is traditionally used.

The second class of methods is traditionally used to determine the most significant factor characteristics that are subsequently used to obtain predictive estimates of other external characteristics or to classify the objects of analysis. Here are traditionally used: canonical correlation analysis (CCA) [9, 10], discriminant analysis (DA) [11, 12], and regression analysis [13, 14]. In a number of cases, direct estimates of the informative value of the indicators are proposed on a basis of a posteriori data [15].

The array of the examined approaches is rather heterogeneous. Often for the same task of determining the most informative attributes, different methods provide different results. At the same time, the question remains open: whether the drift of results is conditioned by the imperfection of methods or whether it reflects the specific features of a particular problem. The proposed research considers a generalization of a number of approximation methods of the initial evaluation matrix with the interpretation of the drift of the results.

2. Analysis method of the information significance of indicators based on the approximation of the evaluation matrix

The task of analyzing the informational significance of the indicators is as follows: let there be given a system (set) of indicators $P = \{p_1, p_2, \cdots p_n\}$, characterizing the analysis objects $O = \{o_1, o_2, \cdots o_m\}$. Without loss of generality, we can consider all indicators of the set to be unidirectional in improving the set of objects. This means that the trajectory of the development of the system of objects leads to progress (improvement) if at the time $t_1 > t_0$ the system of inequalities $\forall\, i = \overline{1,m}, j = \overline{1,n}: p_{ij}(t_0) \leq p_{ij}(t_1)$ is fulfilled, and at least one indicator shows a strict inequality. It is required to choose some minimal subset $B \subset P$ of the most significant indicators in some sense preserving or allowing to recover the greatest amount of initial information from the system of parameters P.

Let the information about the objects of analysis O, characterized by the set of indicators P, be represented by the evaluation matrix:

$$A = \begin{pmatrix} p_{1,1} & p_{1,2} & \cdots & p_{1,n} \\ p_{2,1} & p_{2,2} & \cdots & p_{2,n} \\ \cdots & \cdots & \cdots & \cdots \\ p_{m,1} & p_{m,2} & \cdots & p_{m,n} \end{pmatrix} \tag{1}$$

In the research [16] the author proposed a new approach to the solution of the problem of group selection, based on the use of different matrix norms. This approach can be extended to the task of selecting the most informative indicators.

For this, first of all, let us perform the normalization of the matrix A. Likewise [16] rank normalization will be examined by replacing all indicator values p_{ij} with their ranks. Thus we proceed from matrix (1) to matrix A':

$$A' = \begin{pmatrix} r_{1,1} & r_{1,2} & \cdots & r_{1,n} \\ r_{2,1} & r_{2,2} & \cdots & r_{2,n} \\ \cdots & \cdots & \cdots & \cdots \\ r_{m,1} & r_{m,2} & \cdots & r_{m,n} \end{pmatrix} \tag{2}$$

Let us introduce the matrix of a given rank B, approximating the matrix A'. In order to do this, we proceed from the general idea that with the help of the most informative indicators we can perform a more accurate approximation. If the matrix B has rank 1, then, as an approximation, we consider the matrix B, all columns of which are identical and equal to a column of the A'. Let us denote such a matrix by $B^{\langle i \rangle}, i = \overline{1,n}$, where i – is the selected column of the matrix A'.

To select this column, we consider a function $f(A', B^{\langle i\rangle})$, or which we obtain a minimum of a matrix norm:

$$\min\{\|f(A', B^{\langle i\rangle})\|: i = \overline{1,n}\} \quad (3)$$

As a matrix norm further we shall consider element-wise matrix norm L_p, and as a function $f(A', B^{\langle i\rangle})$ we shall use matrices difference function $f(A', B^{\langle i\rangle}) = A' - B^{\langle i\rangle}$:

$$L_p(f(A', B^{\langle i\rangle}) = \left(\sum_{k=1}^{m}\sum_{j=1}^{n}|r_{kj} - r_{ki}|^p\right)^{1/p} \quad (4)$$

In general it is possible to consider other matrix norms (for example, the Schatten norm S_p). As a function $f(A', B^{\langle i\rangle})$ it is as well possible to use [16] the difference of covariance matrices:

$$f(A', B^{\langle i\rangle}) = A' \cdot (A')^T - B^{\langle i\rangle} \cdot (B^{\langle i\rangle})^T$$

By ranging columns $i = \overline{1,n}$ by matrix norm value (4), we get the ranging of each separate indicator by information significance (the lower value of the matrix norm corresponds to the greater information importance). In this case it is only necessary to go through n options. The parameter of matrix norm p, as shown in [16], reflects ration to risk: at larger value of p large deviations $|r_{kj} - r_{ki}|$ get larger penalties. Thus by changing the parameter p, it is possible to see the drift of priorities of indicators of information significance in relation to risk to get significant discrepancies with at least one of the objects.

With increasing rank of selecting matrix B till the value s from matrix (2) it is necessary to choose s columns. Let us denote such a matrix by $B^{\langle i_1, i_2, \cdots i_s\rangle}$.

For the element-wise matrix norm, the selection of the matrix $B^{\langle i_1, i_2, \cdots i_s\rangle}$ is simplified by the fact that in order to find the minimum of the matrix norm $L_p(f(A', B^{\langle i_1, i_2, \cdots i_s\rangle})$ one can search for a minimum separately for each column h of the matrix A':

$$\sum_{j=1}^{m}|r_{jh} - r_{jk}|^p \to \min$$

where p - the parameter of the element-wise matrix norm L_p, $k \in \{i_1, i_2, \cdots, i_s\}$ is the set of selected columns of the matrix A' (variable selection parameter). If $\mathrm{Rank}(B^{\langle i_1, i_2, \cdots i_s\rangle}) = s$, then it is required to sort out $\binom{n}{s}$ sets $\{j_1, j_2, \cdots, j_s\}$ (for large rank it is reasonable to perform a random search).

Let us note that within the proposed approach it is easy to combine the two classes of autoinformativity and external informativity problems. In the conditions of the criterion of external informativeness, it is necessary to assign a ranking of the objects external to the evaluation matrix. These can be expert rankings or rankings obtained by some methods on the basis of the estimated matrix A. In this case, the ranking can be either fully consistent or contradictory. Let D be a matrix of dimension $m \cdot r$ such rankings from r experts or external ranking methods. Then the task of evaluating the external informativeness of the characteristics consists in choosing a subset of such columns of the estimated matrix A' and forming from them a matrix B, which will provide the minimum element-wise matrix norm $\|D - B\|_p \to \min$.

We also note that in the matrix B, as in the case of the autoinformativity criterion, and in the case of the criterion of external informativity, the chosen indicators can meet several times. This quantity can be a priori considered as the importance of the exponent relative to other exponents of the matrix B. From the primary indicators of matrix A' you can build derived indicators by adding them to the initial set of indicators and evaluating their informativeness.

Finally, in the proposed method, it is possible to combine the autoinformativity criterion with the criterion of external informativity, for this it is sufficient to use the method on combining the columns of the matrices D, A'.

IOP Conf. Series: Journal of Physics: Conf. Series **1084** (2018) 012011 doi:10.1088/1742-6596/1084/1/012011

3. Substantiation of the proposed approach for analysis of the information significance of indicators

To substantiate the proposed approach for assessing the informativeness of indicators, a number of interrelated statistical indicators were modeled and method performance on a priori known relationships was considered. Standard methods were as well used for comparison to analyze the relationship of a number of characteristics based on the principal component method, and the construction of regression dependencies. In the modeling, some indicators were taken independent, other indicators were a function (linear or nonlinear) of independent indicators with the addition of noise:

- random variable normally distributed with the expected value of 12 units, standard deviation 1 unit $p_1 \sim N(12,1)$;
- random variable normally distributed with the expected value of 10 units, standard deviation 3 unit $p_2 \sim N(10,3)$;
- random variable normally distributed within the range of 1 to 5 units $p_3 \sim U(1,5)$;
- random variable distributed by Weibull distribution with scale parameter $\lambda = 1$ and shape parameter $k = 1$: $p_4 \sim W(1,1)$;
- random variable exponentially distributed with rate parameter $\lambda = 10$: $p_5 \sim Exp(10)$;
- random variable $p_6 \sim 5p_1 + 7p_2 + N(0,1)$;
- random variable $p_7 \sim 3p_1 + p_1p_2 + N(0,1)$;
- random variable $p_8 \sim 4{p_3}^3 + p_1p_2 + N(0,1)$;
- random variable $p_9 \sim p_4p_5 + N(0,1)$;
- random variable $p_{10} \sim p_1p_2p_3 + N(0,1)$.

For modeling, the statistical computation environment R was used, for the repetition, a given pseudo-random sequence was chosen. Program code for generating the source data is presented below.

```
N<-100
set.seed(777)
p1<- rank(rnorm(N,12,1))
p2<- rank(rnorm(N,10,3))
p3<- rank(runif(N,1,5))
p4<- rank(rweibull(N,4,20))
p5<-rank(rexp(N,10))

p6<-rank(5*p1+7*p2+rnorm(N,0,1))
p7<-rank(3*p1^2+p2*p1+rnorm(N,0,1))
p8<-rank(4*p3^3+p1*p2+rnorm(N,0,1))
p9<-rank(p4*p5+rnorm(N,0,1))
p10<-rank(p1*p2*p3+rnorm(N,0,1))

Quest <- data.frame(p1,p2,p3,p4,p5,p6,p7,p8,p9,p10)
```

In total $N = 100$ copies of each random variable were generated, which allowed forming an estimated matrix of size $100 \cdot 10$ (100 objects, each of which is characterized by 10 indicators). For the comparability of methods in the future, the normalized values of these indicators were used, the rationing was carried out by ranks. Thus, the task was to select the most informative signs set for the objects, characterized by 10 signs.

Initially, the correlation of the obtained indices was evaluated. Figure 1 shows the cross-correlation between the initial indicators.

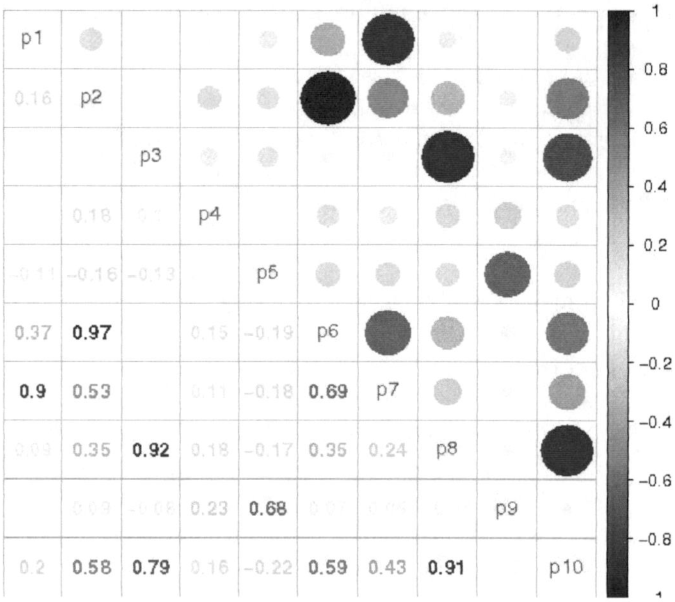

Figure 1. Correlogram of initial indicators (Spearman's rank correlation coefficient is used).

It appears from Figure 1 that independent indicators are characterized by:

- indicator p_1 has strong correlation relationship with p_7, moderate relationship with p_2, p_6;
- indicator $p_2$has strong correlation relationship with p_6 and moderate relationship with p_7, p_{10};
- indicator p_3 has strong correlation relationship with p_8, p_{10};
- indicator p_4 does not have significant relationship with other indicators;
- indicator p_5 has moderate relationship with p_9.

Altogether, the revealed correlation dependencies reflect the above models of constructing indicators. The results of revealing the most informative subset of indicators from the proposed 10 indicators using the developed method are presented below. When selecting one most informative indicator, all 10 indicators were enumerated at different values of the element-wise matrix norm p=2,3,4,10,100. Calculations have shown that with the increase in the indicator p, ranking of indicators by informativeness in the choice of one indicator does not change. Figure 2 shows the deviations of the matrix approximated by one index from the evaluation matrix. It appears from Figure 2 that the most informative indicator is p_{10}, second most informative is p_6; the least informative are p_5, p_9. In general, the estimates obtained correspond to the correlation estimate: indicator p_{10} has maximum correlation relationships with other indicators, indicator p_6 has strongest correlation relationship with p_2.

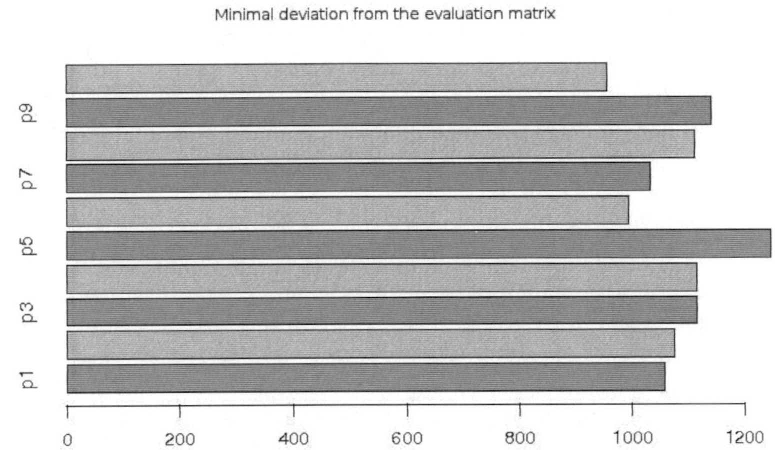

Figure 2. The informative value when choosing one indicator.

During the analysis of the information value of subsets of two indicators, a complete enumeration of 45 options must be performed. The results of such enumeration at the parameter of matrix norm p=2 are shown in Figure 3.

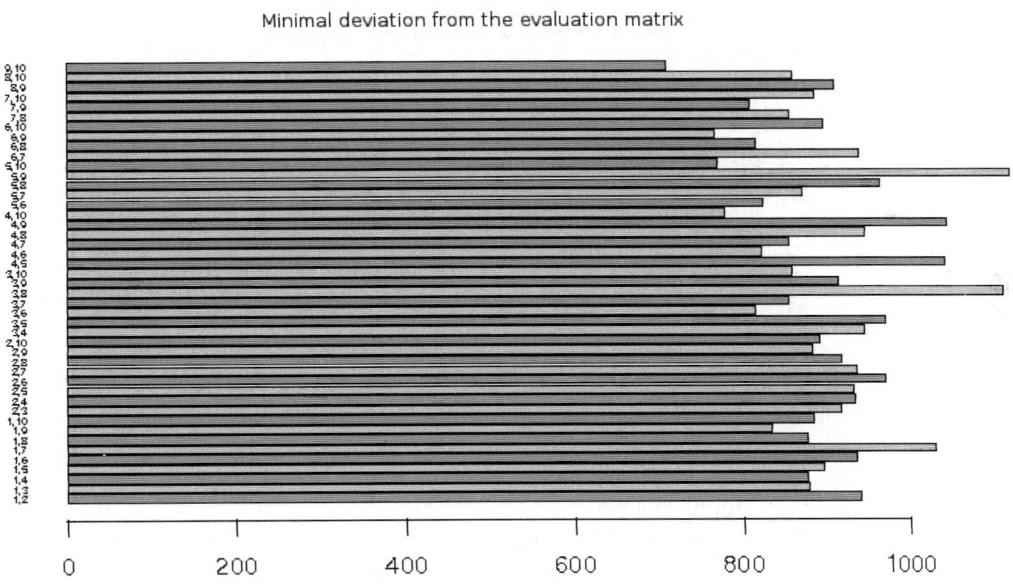

Figure 3. The informative value when choosing two indicators.

It appears from Figure 3 that the most informative set of two indicators is the set of p_9, p_{10}, closest sets to them are: p_6, p_9, as well as p_5, p_{10}. The least informative sets are p_5, p_9, as well as p_3, p_8. In the obtained results it should be noted that in the most informative set of the two indicators, the most informative set of one indicator p_{10}) was combined with the least informative set of one indicator p_9). Considering the structure of the indicator models it is clear that in the set p_9, p_{10} all five independent indices are combined p_1, p_2, p_3, p_4, p_5, of which others are derivatives. Calculations also showed that an increase in the parameter of the matrix norm p smoothes out differences in the informativeness of the indicator sets, but generally does not change the qualitative picture.

Further expansion of the set of indicators leads to a rapid increase in the number of options. For practical purposes, in order to search for an acceptable solution, a random search was implemented. At selecting the three indicators, the most informative set became p_3, p_6, p_9, as well as the set p_6, p_8, p_9. Thus, it is evident again that to one of the most informative sets of two indicators the indicator from the least informative set of two indicators is added: either p_3, or p_8. In addition, the analysis of the structure of models also indicates that by sets p_3, p_6, p_9, or p_6, p_8, p_9 all independent indicators of the array of indicators of interest are covered.

To compare the obtained results the existing methods were used. Figure 4 shows absolute values of the factor loadings of each of the characteristics when the estimated matrix is decomposed into principal components (the first principal component is taken). As the figure shows, the greatest factor loading is observed in indicators p_{10}, p_6, the least factor loading is observed in indicators p_9, p_4. These results almost coincide with the isolation of the most informative indicators by the developed method at one indicator in the approximating subset. However, in this case the developed method does not repeat completely the principal component method (for example, in the developed method the worst-case in approximation of the estimated matrix is the indicator p_5).

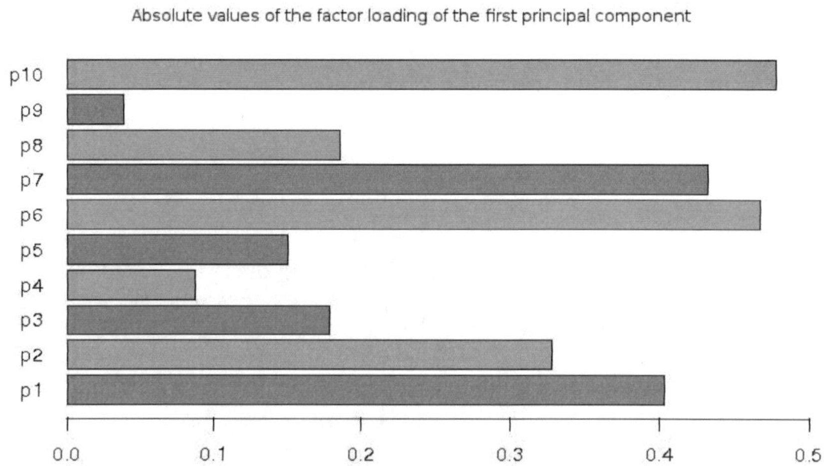

Figure 4. Decomposition by first principal component.

The further use of the principal component method in order to isolate the most informative indices from more than one element becomes complicated, since in the principal component method each main component represented as a linear combination of the sought parameters retains a certain portion of the information of the estimated matrix proportional to the eigenvalue of the principal component. Thus, the developed method extends the method of principal components in this direction.

An important criterion for the information significance of indicators is the construction of the most accurate regression dependencies on their basis. To perform such a comparative evaluation, all possible subsets of the three indicators were analyzed (their number amounted to 120 options). The calculations were carried out in R environment. The program code for such enumeration is presented below.

```
count<-1
main_matrix<-matrix(0,nrow = 120, ncol=5)

for (i in c(1:8))
{
 for (j in c((i+1):9))
 {
  for (k in c((j+1):10))
```

IOP Conf. Series: Journal of Physics: Conf. Series **1084** (2018) 012011 doi:10.1088/1742-6596/1084/1/012011

```
    {
     X<-Quest[c(i)]
     Y<-Quest[c(j)]
     Z<-Quest[c(k)]
     s<-c()
     for (l in setdiff(c(1:10), c(i,j,k)))
     {
      mydata<-data.frame(X,Y,Z,Quest[c(l)])
      my_model<- lm(
       as.formula(paste(colnames(mydata)[4], "~",
       paste(colnames(mydata)[c(1,2,3)],
       collapse = "+"),
       sep = "")),data=mydata)
      s<-c(s, summary(my_model)$r.squared)
     }
     main_matrix[count,1]<-i
     main_matrix[count,2]<-j
     main_matrix[count,3]<-k
     main_matrix[count,4]<-mean(s)
     main_matrix[count,5]<-sd(s)
     count<-count+1
    }
  }
}
write.csv(main_matrix, file = "~/inform.csv", row.names=TRUE)
```

For the evaluation, linear regression dependencies of the performance indicators (all indicators that were not included in the subset of the most informative ones were used as such) from factor characteristics (the indicators, considered the most informative were taken as such) were constructed. Each subset of the most informative features for each obtained model was compared with the coefficient of determination of the model. Eventually, each subset of characteristics taken as the most informative was compared with the average value of the coefficient of determination for all models, as well as the standard deviation of the determination coefficient for all the models built.

Table 1. Analysis of the informativeness of indicators in view of prognosis composition.

p_i	p_j	p_k	mean(R^2)	sd(R^2)
3	6	9	0.6509213449	0.2000243582
6	8	9	0.6495901871	0.1989376844
3	7	9	0.6168643024	0.3155501039
7	8	9	0.6163395132	0.3149012883
3	9	10	0.6120201159	0.2022198237
8	9	10	0.6064224195	0.2029571474
1	3	9	0.582801548	0.334085641
3	4	6	0.5826783665	0.300481417
1	8	9	0.5826300283	0.333471353

Table 1 shows a fragment of the calculation results, which are sorted in descending order of the average determination coefficient. Out of the obtained results it follows that the chosen by the developed methods subsets of the most informative indicators p_6, p_8, p_9, as well as p_3, p_6, p_9 have a maximum average coefficient of determination. Thus, the subset proposed by the developed method, in addition to the features considered, tends to maximize the predictive power of informative indicators, providing as a whole more accurate linear approximation of other indices.

4. Approbation of the method of analyzing the informativeness of indicators

Let us demonstrate the use of the proposed method on the example of the analysis of the information significance of the indicators of the performance of scientific organizations. Estimates were made based on data monitoring for the element-wise matrix norm for different values of the parameter p. The analysis showed that for small values of p preference is given mainly to the financial indicator unit. So, for p = 1 the most informative are: 6_3_3 - The share of internal costs for research and development of funds received to fulfill orders of economic entities; 6_1_3 - Share of income from the rent of machinery and equipment for rent in the total income of the scientific organization (%); 5_1_7 - Average annual cost of fixed assets of research and development per one employee of a scientific organization; 5_1_10 - Average monthly salary of researchers of a scientific organization; 3_1_1 - The volume of funds received from the transfer of technology, attributed to the number of workers in the scientific organization; 1_1_3 - The number of doctoral and candidate dissertations defended during the evaluation period by employees of the scientific organization, referred to the number of researchers.

The least informative in this case are: 6_2_1 - - The share of expenses from ordinary activities in the total actual expenditures of the scientific organization; 5_1_5 - - Share of areas not leased out in the total area of buildings; 1_1_1 - The share of internal costs for research and development in the total volume of works performed by the scientific organization, services.

At p = 2 the priorities between the indicators are somewhat equalized. And the priorities are shifted to the area of the scientific component. So the most informative are recognized: 5_1_2 - Share in fixed assets: machinery and equipment; 4_1_3 - Number of researchers under the age of 39 years, referred to the number of researchers; 1_2_5 - Impact factor of publications of scientists in the Web of Science; 4_1_1 - The proportion of researchers in the total number of workers in a scientific organization; 3_3_1 - List of elements of the innovation infrastructure created by the scientific organization or with its participation.

At the same time, the least informative are: 5_1_5 - Share of areas not leased in the total area of buildings (preserved from p = 1); 1_1_4 - List of state and international awards, prizes, rewards, honorary titles received by a scientific organization; 2_2_1 - The number of scientific and educational structures (departments, laboratories, scientific and educational centers, etc.), created in cooperation with higher educational institutions.

Finally, at p = 10, priorities are gradually shifted to the assessment of the human, technological and publication component of the scientific organization: 4_1_1 (preserved from p = 2), 5_1_2 (preserved from p = 2); 4_1_2 - Number of highly qualified specialists (candidates and doctors of science); 4_1_3 - Number of researchers under the age of 39 years, referred to the number of researchers; 5_1_6 - The share of machinery and equipment up to 3 years of age, inclusive, in the total cost of machinery and equipment; 1_2_6 - Number of published reports, abstracts of papers submitted by scientists at major conferences, symposiums and readings, attributed to the number of researchers; 1_2_5 - Impact factor of publications of scientists in the Web of Science. At the same time, most of the financial indicators accepted as the most priority become the least priority for p = 1: 6_3_3; 6_1_3; 5_1_7; 5_1_10.

Thus we shall conclude that for p = 1 and p = 10, two diametrically opposite views on the informational significance of the indicators are obtained: an increase in p contributes to giving preference to primary driving factors (human, technological); a decrease in p shifts priorities to the area of stimulation of primary factors (income level).

5. Conclusions

The analysis, allowed to reveal the features of the proposed method in comparison with the existing approaches, is presented in Table 2.

Table 2. Advantages and disadvantages of the proposed method.

Property	The developed method	PCA	Regression
Analysis of autoinformativity	+	+	-
Analysis of information significance in relation to external features	+	-	+
Analysis of information significance of one indicator	+	+	+
Analysis of information significance of group of indicators	+	-	+
Analysis of the drift of priorities of information significance	+	-	-
Presence of an effective computational algorithm	-	+	+

The analysis of the correlation relationships of the indicators made it possible to conclude that the developed method is oriented towards indicators that have the maximum number of the strongest statistical links with the rest of the indicators.

Analysis of data by the method of principal components made it possible to conclude that the developed method tends to select a subset of informative indicators, minimizing the information loss of the initial evaluation matrix, on one indicator the developed method and the method of principal components practically coincide; however, for the developed method, it is possible to expand a subset of informative indicators without a linear aggregation.

Analysis of regression dependencies on various subsets of indicators made it possible to conclude that using the developed method on the selected subset of informative indicators the construction of the most accurate linear regression dependencies is possible.

References

[1] Rico-Sulayes A 2017 Reducing Vector Space Dimensionality in Automatic Classification for Authorship Attribution *Revista Ingeniería Electrónica, Automática y Comunicaciones* **38 (3)** 26–35

[2] Wang Y, Liu Y and Wu L 2011 Study on background modeling method based on robust principal component analysis *Annual Conference on Electrical and Control Engineering ICECE* 6787–6790

[3] De La Torre F and Black M 2003 A framework for robust subspace learning *International Journal on Computer Vision* 117–142

[4] Eckart C and Young G 1936 The approximation of one matrix by another of lower rank *Psychometrika* **1 (3)** 211–218

[5] Gillis N 2012 Sparse and unique nonnegative matrix factorization through data preprocessing *Journal of Machine Learning Research* **13** 3349–3386

[6] Liu J, Wang C, Gao J and Han J 2013 Multi-View Clustering via Joint Nonnegative Matrix Factorization *Proceedings of SIAM Data Mining Conference* 252–260

[7] Zhang T and Golub G H, 2001 Rank-one approximation to high-order tensors *SIAM J Matrix Analys* **23** 534-550

[8] Tyrtyshnikov E 2003 Tensor Approximations of Matrices Generated by Asymptotically Smooth Functions *Sb. Math.* **194:6** 147-160

IOP Conf. Series: Journal of Physics: Conf. Series **1084** (2018) 012011 doi:10.1088/1742-6596/1084/1/012011

[9] Hotelling H 1936 Relations Between Two Sets of Variates *Biometrika* **28 (3–4)** 321–377

[10] Hardoon D R, Szedmak S and Shawe-Taylor J 2004 Canonical correlation analysis: an overview with application to learning methods *Neural Computation* **16** 2639–2664.

[11] Fisher R A 1936 The Use of Multiple Measurements in Taxonomic Problems *Annals of Eugenics* **7 (2)** 179–188

[12] Asghari O and Hezarkhani A 2008 Applying discriminant analysis to separate the alteration zones within the Sungun porphyry copper deposit *Journal of Applied Sciences* **24** 4472-4486

[13] Fornalski K W 2015 Applications of the robust Bayesian regression analysis *International Journal of Society Systems Science* **7 (4)** 314–333

[14] Andersen R 2008 Modern Methods for Robust Regression *Sage University Paper Series on Quantitative Applications in the Social Sciences* 07-152

[15] Siddiqi N 2015 SAS Institute Inc., Cary, North Carolin, *Credit Risk Scorecards: Developing and Implementing Intelligent Credit Scoring* 208 p

[16] Artamonov Yu 2016 Group Selection Using Matrix Norms *Izv. Irkutsk State University. University. Ser. Mathematics* **18** 3-20

IOP Conf. Series: Journal of Physics: Conf. Series **1084** (2018) 012012 doi:10.1088/1742-6596/1084/1/012012

Features of diffuse photon migration in soft biological tissue

A Potlov, S Frolov[1] and S Proskurin

Biomedical Engineering, Tambov State Technical University, 392000, Sovetskaya 106, Tambov, Russia

[1]E-mail: sergej.frolov@gmail.com

Abstract. Specific features of photon density normalized maximum movement in soft biological tissue under the influence of deforming force are described. Soft biological tissue is simulated as a turbid and linear isotropic pseudo-incompressible medium. It has been shown that the deforming effects do not influence the direction of photon density normalized maximum (PDNM) movement, but affect the speed of its movement. The decay of diffuse transmittance intensity is 8 to 10% faster in undeformed state than in those with deformations. The established features can be useful for the generation of an initial approximation to the spatial distributions of optical parameters during solution of the diffuse optical tomography (DOT) inverse problem.

1. Introduction

Optical radiation scattering through biological object gives useful information about tissue structure. However, due to strong scattering and significant nonlinearity of photon trajectories there are difficulties in interpretation of experimental results (aquired TPSF – time point spread function) and challenges in image reconstruction in diffuse optical tomography (DOT). Overcoming the difficulties is possible by using numerical simulation of radiation passing through human-tissue-like phantom which migration is of extreme importance [1-3]. Among features the solution of the inverse problem of DOT (image reconstruction) is most often reduced to a multiple iterative solution of the forward problem (numerical simulation of photon migration) with the calculation and minimization of the discrepancy between the TPSFs obtained as a result of numerical simulation and the TPSFs from detectors of DOT system [4, 5].

It is known, that tissue external shape can significantly affect the quality of DOT image reconstruction [4, 6]. Therefore, it is necessary to provide good tight contact of optical fibers to inject and to detect diffusely transmitted photons in the investigated area. It can significantly reduce the loss of useful signal, and thus to improve the accuracy of absorption and scattering properties mapping [5, 7]. However, in some cases, this also leads to the deformation of tissue shape. In this regard, compression plates are usually appears as the cause of the deformation [5, 6]. In authors research is used an elastic bracelet to fixture the irradiation source and detector fibers, which allows achieving good contact of optical ports with the skin at acceptable deformations of individual parts of the investigated object. Application leads to the image artifacts caused by the discrepancy between actual boundaries of investigated area and its mathematical model representation [6-8]. Due to the susceptibility to deformation these artifacts can complicate examination of a soft tissue like breast [3, 7].

The purpose of this work is to identify specific features of the photon density normalized maximum (PDNM) movement in biological soft tissues with considering possible tissue deformation. The influence of tissue deformations on photon density distribution is numerically simulated using diffusion approximation to the radiative transfer equation (optical properties simulation) and basic equation of the elasticity theory for quasi-static deformation (stress-related properties simulation).

2. Experimental part

2.1. Light propagation model

To describe the diffusion of photons in biological tissue we used particular case of diffusion approximation of the radiative transfer equation. It was the Model of a Dropusing – radiation pulse containing fixed initial number of photons that appeared in the object near its surface and diffused within the object [8]. Such approach allows sufficiently precise description of experimental data for homogeneous and inhomogeneous cases and it is based on the solution of the radiative transfer equation for a light pulse, containing a finite number of photons. According to the diffusion equation, the density of diffusing photons is described as [9-11]:

$$c^{-1}\frac{\partial\varphi(x,y,z,t)}{\partial t}-D(x,y,z)\nabla^2\varphi(x,y,z,t)+\mu_a(x,y,z)\varphi(x,y,z,t)=S(x,y,z,t),\ \forall x,y,z\in\Omega, \quad (1)$$

where $c=c_0/\nu_{\text{object}}$ is light speed in the medium; c_0 – light speed in vacuum; ν_{object} –relative refractive index of the simulated object (Ω) and its boundaries ($\partial\Omega$); x,y,z are coordinates of all points of the final simulated area consisting of inner part of simulated object Ω, its boundaries $\partial\Omega$, radiation source (q), detectors, and the medium surrounding the object;

$$D(x,y,z)=\left\{3\left[\mu_a(x,y,z)+\left(1-g(x,y,z)\right)\mu_s(x,y,z)\right]\right\}^{-1} \quad (2)$$

where D is the diffusion coefficient at the position x,y,z; $\mu_a(x,y,z)$ is diffusion and absorption coefficients; $\mu_S(x,y,z)$ –scattering coefficient; – anisotropy factor (the average cosine of the scattering angle); $\phi(x,y,z,t)$ –photon density at the point with coordinates x,y,z at a time t; and $S(x,y,z,t)$ – photon source function.

The boundary condition of the third kind (the Robin condition) was used to describe the photon flux on the object boundary. The flux of photons, leaving the finite domain through the boundary, was equal to the flux at the boundary, multiplied by the coefficient including the light back reflection into the object [11-13]:

$$\varphi(x,y,z,t)+2D(x,y,z)F\frac{\partial\varphi(x,y,z,t)}{\partial\hat{n}(x,y,z)}=0,\ \forall x,y,z\in\partial\Omega,\ x,y,z\notin q, \quad (3)$$

where $\hat{n}(x,y,z)$ is direction of the outer normal to the boundary $\partial\Omega$ at the point with coordinates x,y,z. F – Fresnel reflection coefficient [4, 14] was calculated as follows:

$$F=\frac{2/(1-R_0)-1+\left|\cos(Q_c)\right|^3}{1-\left|\cos(Q_c)\right|^2}, \quad (4)$$

where R_0 and Q_c – coefficients, respectively equal to:

$$R_0=\left(\frac{\nu_{\text{object}}}{\nu_{\text{medium}}}-1\right)^2\cdot\left(\frac{\nu_{\text{object}}}{\nu_{\text{medium}}}+1\right)^{-2} \text{ and } Q_c=\arcsin\left(\nu_{\text{medium}}/\nu_{\text{object}}\right), \quad (5)$$

where ν_{medium} – the relative refractive index of the medium surrounding the object (for the air $\nu_{\text{medium}}=1$).

After completion of the simulation iterative process PDNM function, $\varphi(x,y,z,t)$, is normalized [3] with respect to its maximum $\varphi_{\max}(x,y,z,t)$:

$$\varphi_{norm}(x,y,z,t) = \varphi(x,y,z,t)/\varphi_{\max}(x,y,z,t), \tag{6}$$

and represented as follows:

$$\varphi_{PDNM}(x,y,z,t) = \begin{vmatrix} 1, \ \varphi_{norm}(x,y,z,t) \geq P \\ \varphi_{norm}(x,y,z,t), \ else \end{vmatrix}, \tag{7}$$

where P – is the experimentally determined minimum of photon density level $0 < P \leq 1$.

2.2. Breast deformation model

Soft biological tissues are nonlinear elastic media. Nevertheless, in some cases, for example when the deforming forces cause small bending ($\leq 5\%$), the soft tissues can still be considered as a media with linear properties. For example in DOT, the breast tissue should be presented as a linear isotropic pseudo-incompressible medium [3, 5-7].

In this case, the basic equation of the elasticity theory for quasi-static deformation on the internal nodes of the simulated area is given by [3, 15]:

$$(\lambda + \mu)\nabla(\nabla u) + \mu\nabla^2 u = 0, \tag{8}$$

and the same equation for nodes on the boundary $\partial\Omega$ of the studied area is represented as follows [6]:

$$((\lambda + \mu)\nabla(\nabla u) + \mu\nabla^2 u)\cdot \breve{n} = n, \tag{9}$$

where $\breve{n}$ – a unit vector directed outwards from Ω; h – represents the tension on the surface and boundary of the simulated area; $u = (u_1, u_2, u_3)$ – the displacement vector components at the axes x, y, z in the Cartesian coordinate system; μ and λ – Lame's elastic constants. These constants for isotropic medium (first and second Lame's elastic constants) are associated with the Young's modulus E and Poisson ratio, ν, as follows [7, 16]:

$$\mu = 0.5\cdot E\cdot(1+\nu)^{-1} \tag{10}$$

and

$$\lambda = \nu E\cdot\left[(1+\nu)(1-2\nu)\right]^{-1}. \tag{11}$$

Thus, the simulated object was considered to be free of any initial deformations and an influence of the internal deforming forces (for example, muscle activity). All deformations were considered to be caused by external loads as source and detection fibers and their fixtures [3].

To simplify the calculations it is assumed, that these fibers are located on slightly reflective elastic band with adjustable diameter consisting of two identical halves. It allows fixing painlessly the source and detection fibers on the investigated biomedical object. Only one fiber was used for the injection of photons, meanwhile detection fibers surrounded the investigated object, and were located at the equal angles to each other to the right and left from the source fiber. Thus, the investigated object will be deformed in a plane of the fibers only. Cross-section made at this height will appear as an ellipse rather than a circle. In the simulation of elastic properties of biological tissue it will be assumed that Poisson ratio $\nu = 0.495$, Young's modulus $E = 20$ kPa [3].

It should be noted that scattering and absorption coefficients of the tissue under the deformation may insignificantly change [15, 16], therefore suggested model does not consider these changes.

The described approach was numerically implemented using LabVIEW software package. The numerical solution of the equations (1) - (11) was performed using finite difference method (FDM), using the implicit difference scheme built on seven-point grid pattern [17].

It is important to note that in contrast to the finite element method (FEM) FDM is characterized by a regular grid in all directions, that is more convenient for determining specific features of radiation propagation. Further, compared to Monte-Carlo simulation the diffusion approximation of radiative transfer equation has higher computing speed, that is beneficial for serial experiment.

IOP Conf. Series: Journal of Physics: Conf. Series **1084** (2018) 012012 doi:10.1088/1742-6596/1084/1/012012

We also used numerical simulation to determine the effect of bioobject deformation on the fitures of PDNM movement.

3. Results and discussion

Photon density distribution was consequently simulated (P =0.995-0.999) for the time-resolved cases using pulsed irradiation in homogeneous and inhomogeneous, deformed and undeformed conical objects with optical properties of breast tissue. Figure 1 shows the position of spherical inhomogeneity in a conical object. The optical properties of the single inhomogeneity and the homogeneity part of phantom are chosen in such a way as to correspond to real clinical data.The radius of the heterogeneity corresponds to the average radius of the tumor, taking into account the region of angiogenesis [8, 11] The slice of the object is taken at the half height of the object and in the plane of the fibers [3, 4, 8].

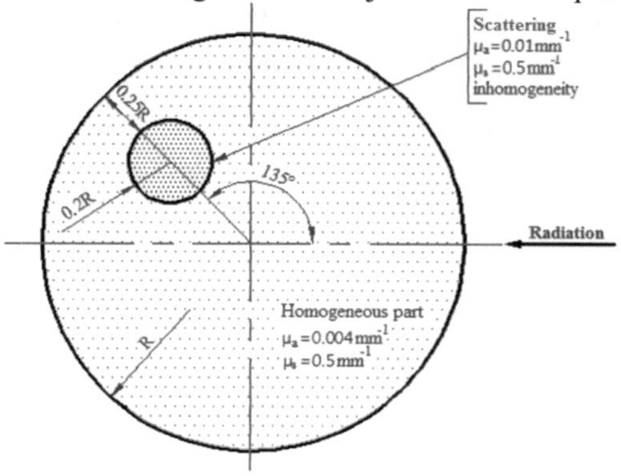

Figure 1. Spatial distribution of optical properties in the plane taken at the half height of heterogeneous conical object (inhomogeneity is considered to be spherical).

Figure 2 shows the simulation results of the photon distribution in the plane (made at the level of the source and detector fibers) of homogeneous undeformed object at different moments of time t Figure 2a corresponds to t=0.7 ns, and Figure 2b to t=3 ns. Optical properties, absorption, $\mu_a(x,y,z)$, and reduced scattering, $\mu'_s(x,y,z)=(1-g(x,y,z))\mu_s(x,y,z)$, coefficients for $\forall(x,y,z)\in\Omega$ are equal to 0.004 mm^{-1} and 0.5 mm^{-1}, respectively [18]. The same simulation results for the optically homogeneous deformed object are shown in Figure 3a and Figure 3b.

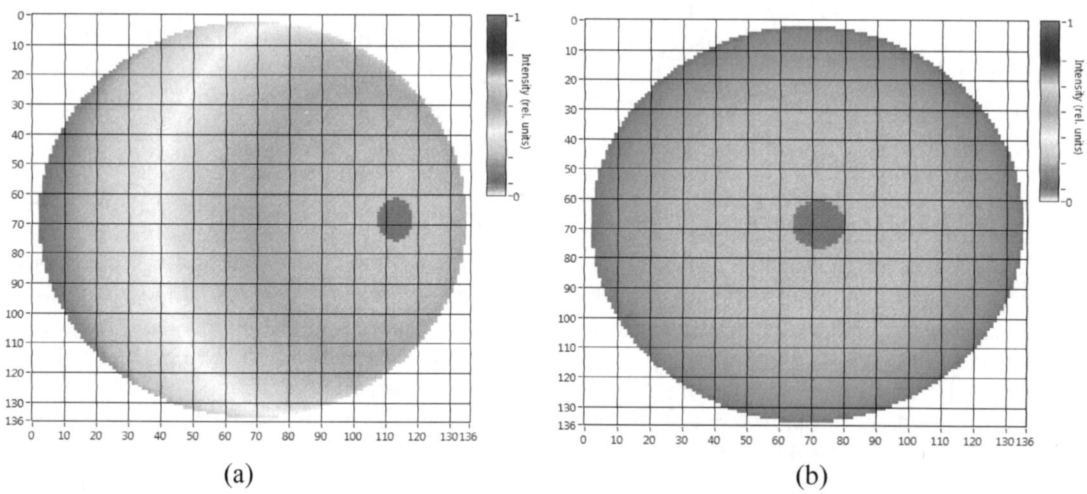

Figure 2. Photon density distributions in the slice of the homogeneous conical object at the following times after the injection of the pulse: 0.7 ns (a) and 3 ns (b).

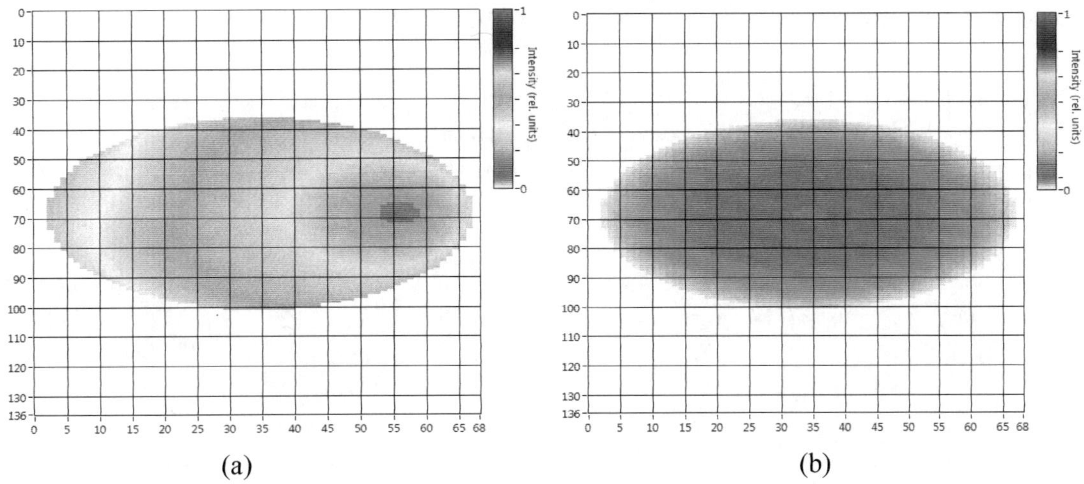

Figure 3. Photon density distributions in the slice of the deformed conical homogeneous object at the following times after the injection of the pulse: 0.7 ns (a) and 3 ns (b). PDNM moves to the center.

Similar simulations have been performed in a same conical object with a spherical absorbing inhomogeneity inside of it [13, 14]. Inhomogeneity diameter is 0.4 of the object radius, located at the angle of 135° with respect to the axis of the incident irradiation, at the depth of 0.25 of the plane radius, R, $\mu_a(x,y,z) = 0.01$ mm^{-1} [18]. Photon density distributions in the plane at the half height of the inhomogeneous undeformed object at time points 0.7 ns and 3.5 ns are presented in Figure 4a and Figure 4b, respectively. As it can be seen in the case of absorbing inhomogeneity, PDNM moves toward the point, which is symmetrical to geometrical center of the heterogeneity relative to the center of the investigated object. Separate investigations have been shown that increasing inhomogeneity absorbing coefficient and size of it leads to increase of PDNM motion speed [14].

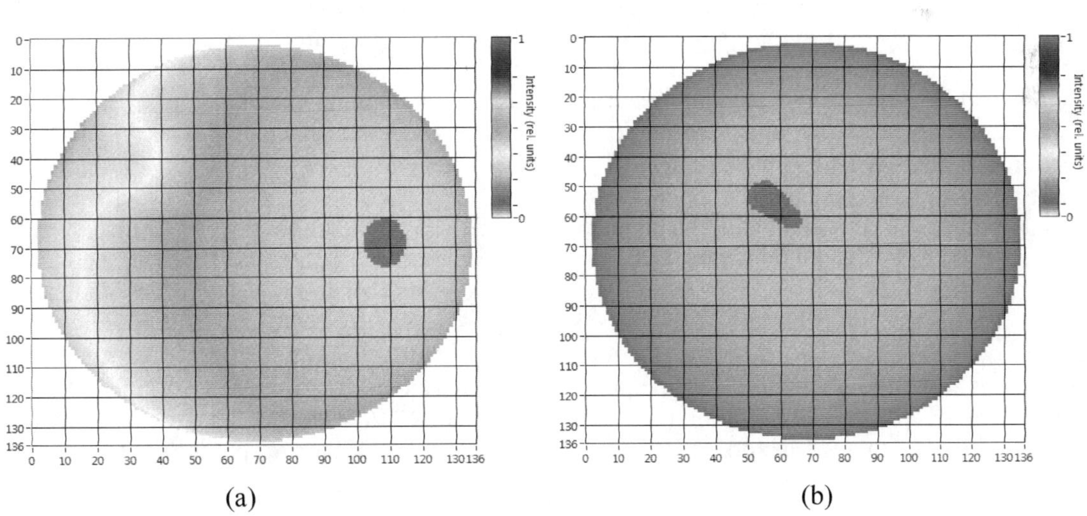

Figure 4. Photon density distributions in the slice of the inhomogeneous conical object at the following times after the injection of the pulse: 0.7 ns (a) and 3.5 ns (b).

For the object in the deformed state the similar results were obtained. Figure 5a (t=0.7 ns) and Figure 5b (t=3.5 ns) shows that PDNM moves similarly, hence the object deformation and its embedded inhomogeneity have no significant influence on the photon density and its normalized maximum [18].

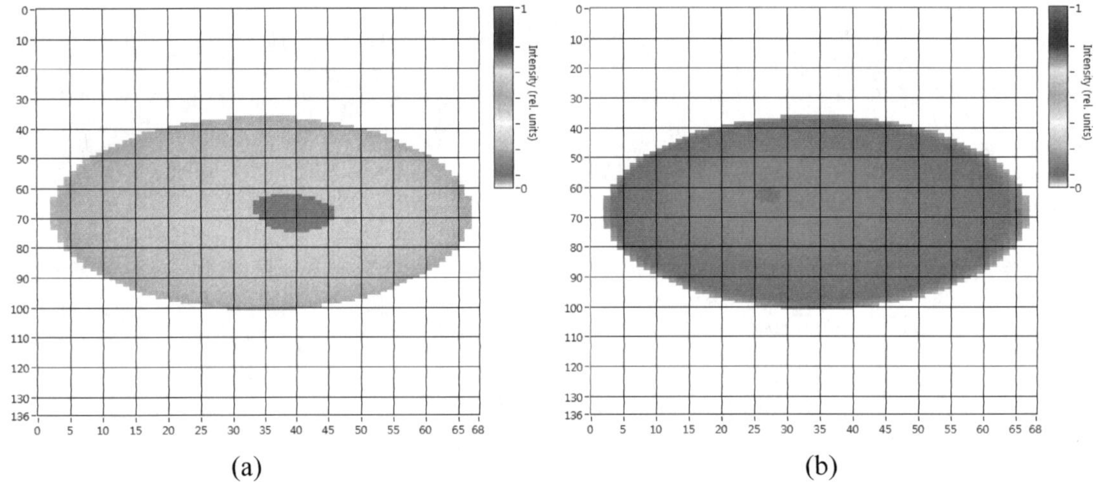

Figure 5. Photon density distributions in the slice of the deformed conical inhomogeneous object at the following times after the injection of the pulse: 0.7 ns (a) and 3.5 ns (b).

However, it has been found that rate of the diffuse transmittance intensity decay in the undeformed objects is slightly higher (8-10%) than in those with deformations. The revealed specific features were verified using tissue-like phantoms [3].

Thus, the deforming effects do not influence the direction of PDNM movement, but they affect the speed of its movement. This means that soft tissue deformations should be taken into account in DOT mammography system design [19] and in solution of the inverse problem based on the acquired data.

In the design of an efective DOT system, glass compression plates (which cause severe deformations of tissue), or the cup-shaped applicator (which has minimum deformation, but poor contact with the investigated biological tissue), should be replaced with an elastic bracelet with injection and detection ports. This approach provides a combination of acceptable deformations and acceptable loss of a useful signal. In addition, an elastic reflective bracelet-holder of injection and detection ports makes it possible to simplify DOT system design.

The software emprovements in reconstruction algorithm should consist of more effective initial approximations to the solution of the inverse problem. If the radiation intensity registered by all detectors of time-resolved DOT system in the region of late arriving photons (LAP) is approximately the same, this should be interpreted as the relative homogeneity of the investigated biological tissue (PDNM reaches the geometric center of the object). If LAP intensity decreases more rapidly in one or more detectors of a region, this can be interpreted as evidence of the presence of an absorbing heterogeneity near to them (PDNM at the point symmetric to the center of the absorbing pathological structure). If the intensity of radiation on one or several detectors decreases slower, then a scattering heterogeneity near these detectors (PDNM in the center of the scattering pathological structure) is highly likely [20].

Conclusions

Numerical models of optical and mechanical properties of biological tissues were described. A series of numerical simulations considering effect of deformations of biological tissues on the trajectory of the PDNM motion are presented. It has been revealed that independently on the values of the absorption and scattering coefficients PDNM moves toward the geometric center of the investigated opticaly homogeneity object in the deformed and undeformed states. Then, in presence of an absorbing inhomogeneity, PDNM moves toward the point being symmetric to its geometric center in relation to the center of conical object regardless with or without deformation. Finally, it has been shown that in the undeformed object radiation intensity decay is about 8-10% higher than in the deformed one.

The significance of the described specific features favour better understanding the influence of different factors on optical radiation propagation and decay in biological tissues. The specific features of PDNM movement in a deformed and undeformed biological object or tissue-like phantom are useful in developing effective methods for solving DOT inverse problem described in [3, 17, 20], since they have more precise initial approximation to spatial distribution of the optical properties in investigated tissue.

An important direction of further development is more accurate accounting for the decrease in average absorbing coefficient of breast tissue in the deformed case This effect is quite important for mammography DOT applications.

Acknowledgements

This work is supported by The Russian Science Foundation (RSF Project 16-15-10327).

References

[1] Patterson M, Chance B and Wilson B 1989 Time resolved reflectance and transmittance for the noninvasive measurement of tissue optical properties *Appl. Optics* **12** 2331

[2] Zimnyakov D A and Tuchin V V 2002 Optical tomography of tissues *Quantum Electron+* **10** 849

[3] Potlov A Yu, Avsievich T I, Frolov S V and Proskurin S G 2016 Forward problem of time-resolved diffuse optical tomography considering biological tissue deformation.chapter of the multi-authored monograph *Computational Biomechanics for Medicine* (Switzerland: Springer International Publishing) p 51

[4] Proskurin S G and Potlov A Yu 2013 Early- and late-arriving photons in diffuse optical tomography *Photon. Las. Med.* **2** 139

[5] Abascal J, Arridge S, Atkinson D, Shindmes R, Fabrizi L, DeLucia M, Horesh L, Bayford R and Holder D 2008 Use of anisotropic modelling in electrical impedance tomography: description of method and preliminary assessment of utility in imaging brain function in the adult human head *NeuroImage* **2** 258

[6] Dehghani H, Srinivasan S, Pogue B and Gibson A 2009 Numerical modelling and image reconstruction in diffuse optical tomography *Phil. Trans. R. Soc. A* **367** 3073

[7] Joldes G R, Wittek A and Miller K 2009 Suite of finite element algorithms for accurate computation of soft tissue deformation for surgical simulation *Med. Image Anal.* **6** 912

[8] Proskurin S G, Potlov A Yu and Frolov S V 2015 Specific features of diffuse photon migration in highly scattering media with optical properties of biological tissues *Quantum Electron+* **6** 540

[9] Tretyakov E V, Shuvalov V V and Shutov I V 2001 Fast approximate statistical nonlinear algorithms for diffusion optical tomography of objects with a complicated internal structure *Quantum Electron+* **12** 1095

[10] Konovalov A B, Vlasov V V, Kalintsev A G, Kravtsenyuk O V and Lyubimov V V 2006 Time-domain diffuse optical tomography using analytic statistical characteristics of photon trajectories *Quantum Electron+* **11** 1048

[11] Arridge S R, Cope M and Delpy D T 1992 The theoretical basis for the determination of optical pathlengths in tissue: temporal and frequency analysis *Phys. Med. Biol.* **7** 1531

[12] Boas D A, Brooks D H, Miller E L, Di Marzio C A, Kilmer M, Gaudette R J and Zhang Q 2001 Imaging the body with diffuse optical tomography *IEEE Signal Processing* **6** 57

[13] Durian D and Rudnick J J 1999 Spatially resolved backscattering: implementation of extrapolation boundary condition and exponential source *Opt. Soc. Am. A.* **4** 837

IOP Conf. Series: Journal of Physics: Conf. Series **1084** (2018) 012012 doi:10.1088/1742-6596/1084/1/012012

[14] Schweiger M and Arridge S R 1999 Application of temporal filters to time resolved data in optical tomography *Phys. Med. Biol.* **44** 1699
[15] Dehghani H, Doyley M M, Pogue B W, Jiang S, Geng J and Paulsen K D 2004 Breast deformation modelling for image reconstruction in near infrared optical tomography *Phys. Med. Biol.* **49** 1131
[16] Pathmanathan P 2006 Predicting tumour location by simulating the deformation of the breast using the finite element method and nonlinear elasticity *PhD thesis* (United Kingdom: University of Oxford) p 217
[17] Potlov A Yu, Frolov S V and Proskurin S G 2015 Movement of the photon density normalized maximum in homogeneous and inhomogeneous media with tissue-like optical properties *Laser Physics* **25** 035601
[18] Konovalov A B, Genina E A and Bashkatov A N 2016 Diffuse optical mammotomography: state-of-the-art and prospects *J. Biom. Photon. Eng.* **2**(2) 020202
[19] Riley J, Amyot F and Pohida T 2012 A hematoma detector - a practical application of instrumental motion as signal in near infra-red imaging *Biomed. Optics Express* **1** 192
[20] Potlov A Yu, Frolov S V and Proskurin S G 2016 Localization of Inhomogeneities in diffuse optical tomography based on late arriving photons *Optics and Spectrosc+* **1(120)** 9

IOP Conf. Series: Journal of Physics: Conf. Series **1084** (2018) 012013 doi:10.1088/1742-6596/1084/1/012013

Aspects of modeling the magnetically controlled thermal accumulators for aerospace engineering

A Shchegolkov[1], A Shchegolkov[2], A Chursin[3] and S Murtuzalieva[4]

[1,2] Tambov State Technical University, 106 Sovetskaya Street, Tambov 392000, Russian Federation

[3,4] Department of Applied Economics of the RUDN University, Miklukho-Maklaya Str. 6, Moscow 17198, Russian Federation

[1]E-mail: energynano@yandex.ru

Abstract. Approaches for creating magnetically controlled thermal accumulators for aerospace engineering based on thermomagnetic convection under the action of a magnetic field created by neodymium magnet under thermal conditions are discussed. The properties of thermal accumulating material (TM) containing magnetic particles are studied. To investigate the distribution of the temperature field it was used a contactless measurement method realized using a thermal imager. The formulation of the problem of mathematical modeling for a thermal control system with TM based on the finite integral transform method is carried out. It allows to study the distribution of temperature fields at design stage. The application of the effect of thermomagnetic convection makes it possible to intensify heat exchange in TM, which increases the efficiency of charge and discharge of TM. This approach is economically feasible and technologically adapted to aerospace technology.

1. Introduction

Today much attention is paid to the creation of "smart" materials that have manageable functional properties and are able to support the given parameters in different conditions. Particular importance in the technology of the "smart" materials creation can take materials that have thermoregulating properties. Thermoregulating materials can be widely used in aviation and space engineering. First of all, this is due to the need of effective thermoregulation of the internal area of aerospace equipment, which is associated with providing comfortable temperature for passengers and service staff, as well as saving the energy resources. Existing technological solutions are associated with the use of electric heaters, which have a decentralized arrangement; while at the desired periods of time there is their heating with thermal energy impact into the internal area, where the passengers and crew are located. Large volumes of internal area and power limitation of thermoregulation system do not allow providing the effective thermoregulation with creation of comfortable conditions. There were carried out studies in the work [1], and it was shown how the duration of airplanes flight affects the temperature modes inside the cabin. The authors found that the short flights had difficulties with creation of comfortable conditions. According to the authors, the solution is an improved model of the temperature mode. In order to obtain objective data related to temperature modes in aircraft cabins, in [2] there were temperature modes measured on mannequins in the full-scale 21-seat section of the aircraft cabin, and the measurements on mannequins were correlated with subjective estimates of the thermal sensation of various parts of the body. The results of the studies [2] showed that local thermal sensations can be predicted using the mannequins.

To increase the efficiency of electric heating systems, there is a potential associated with the implementation of adaptive thermoregulation using thermal accumulators.

Taking into account the large internal volume of the airplanes' cabin, it is necessary to use approaches related to the implementation of multifunctional electric heating systems containing a thermal accumulating buffer, which has adjustable heat conductivity. It will provide a higher sensitivity to the temperature inside the cabin and reduce the consumption of electrical energy for heating. It is reasonable to realize the thermal accumulating buffer using the magnetically controlled thermal accumulators in the mode of thermomagnetic convection [3]. For realization of the thermoregulation system based on thermal accumulating materials, it is necessary to consider technologies of thermal energy accumulating using nanomaterials.

In the article [4] there were shown studies of thermal conductivity of materials with a self-regulation effect in the modified material structure. Self-regulation of "magnetic graphene" improves the thermal conductivity of composites and epoxy resins. When testing the modified material using computer chips, the temperature was reduced by 10 °C compared to other materials. At the same time the authors [4] consider materials that do not have a phase transition, which reduces their efficiency while accumulation of thermal energy. There are questions arises about imparting magnetic properties to nanostructures. Functionalization of carbon nanotubes (CNTs) through magnetic nanoparticles was studied in works [5-6]. The effect of orientation and alignment of CNTs embedded in an epoxy polymer matrix under the influence of a magnetic field on the mechanical properties of obtained nanocomposite was investigated in [5]. The authors found that when alignment of CNTs occurs under magnetic field conditions, the properties of the obtained composites are superior to properties that have not been exposed by magnetic field. In work [6] there are investigated physicochemical mechanisms, under which positive charges ensure effective adsorption of negatively charged magnetic nanoparticles on the graphene surface by means of electrostatic interactions.

In [7], composites of thermoplastic polyurethane elastomer (TPU) / BNNSs (hexagonal boron nitride nanosheets) are made by combining modification and magnetic equalization. In order to impart magnetic properties to BNNSs, it was processed in the manner in which Fe_3O_4 particles were generated on the BNNS surface by oxidizing the aqueous suspension of $Fe(OH)_2$ at room temperature. The authors of the paper obtained thermal conductivity (TPU) / BNNSs at the level of 5.15 W m^{-1} K^{-1} at BNNSs of 30% by weight. In [8] another approach was used to create magnetically controlled 2D nanofiller. In [8], a surface modification of nano-networks of boron nitride (BNN) is used. For this purpose, after the polymerization reaction of BNN, redox polymerization of acrylamide occurs on the surface of already functional hydroxyl-functional BNN. Subsequently, magnetic iron oxide nanoparticles are introduced into the modified BNN by the chelating action of polyacrylamide. Finally, to ensure high-conductivity and magnet-sensitive nanocomposites, the functionalized Fe_3O_4 / BNN is dispersed in an epoxy matrix and aligned under an external magnetic field. Based on the results of thermal conductivity, the thermal transfer of nanocomposites increases to 0.37 W m^{-1} K^{-1}, with only 10% of the functionalized Fe_3O_4 / BNN (at 25 ° C) applied.

So, the main trend in improving the thermophysical properties of composites is associated with the assignment of magnetic properties to nanoscale materials. However, these researches are aimed at studying the properties of solidified composites. At the same time, it is possible to significantly improve heat transfer by using non-hardening materials with phase transitions in which thermomagnetic convection is possible. The problems that should be solved when creating thermoregulation systems with the use of thermal accumulators (TA) based on TM (thermal accumulating material) should include:

- to investigate the properties of thermal accumulating material containing carbon nanostructures with nickel-zinc ferrite;
- to develop the concept of mathematical modeling with the possibility to obtain analytical dependencies, which will allow quick organization of the prototyping the thermoregulation system.

2. Methods and materials

To produce magnetically controlled thermal accumulating material (TM) paraffin (P-2 (Lukoil, Russia)) and modified multi-walled carbon nanotubes (MWCNTs) of the "Taunit" series (NanoTechCenter, Tambov, Russia) were used. MWCNTs were mechanically activated with nickel-zinc ferrite 400HH Ni1-xZnxFe$_2$O$_4$ (NanoTechCenter, Tambov, Russia). Ni1-xZnxFe$_2$O$_4$ was preliminary dispersed to size of 2-4 μm in a ball mill for 60 minutes.

Modifying of paraffin by MWCNTs was performed via ultrasonic radiation and under the exposure of magnetic field (up to 1500 mT). The modifying of paraffin by obtained MWCNTs occurs during the phase transition of paraffin to the liquid state.

The obtained material was granulated in a pressing granulator to form the granules with the size range of 0.5 to 1 mm. After this operation the granules were placed into a carrier fluid. Synthetic motor oil CASTROL W30 (BP, United Kingdom) was used as the carrier fluid. Motor oil and magnetic filler were weighed on the Unit G-200 (UNIT, China) scales with an accuracy of ± 5 mg. After this the filler was introduced into the motor oil and mixed thoroughly. As a result, the liquid phase – motor oil was 20 %, and the solid phase – nanomodified composite – 80%.

The hollow container with the form of a parallelepiped was loaded with magnetically controlled TM. The container from one plane contained neodymium magnet (1000 mT) (NPO Magneton, Russia). Neodymium magnet was made in the form of a cylinder with 50 mm diameter and 10 mm height. The counter plane contained an electric heater. The heating temperature reached to 70 °C.

FLUKE Ti9 thermal imager with 160 x 120 emission receiver, matrix in the focal plane and measurement range of - 20 ° C to + 250 ° C with an accuracy of ± 0.5 ° C was used for contactless measurement of temperature field in the TM.

3. Results of experimental studies and analysis

Figure 1 shows the distribution of the temperature field in the TM taking into account the effect of the magnetic field over time of 45 sec and heating up to 70 °C (thermomagnetic convection).

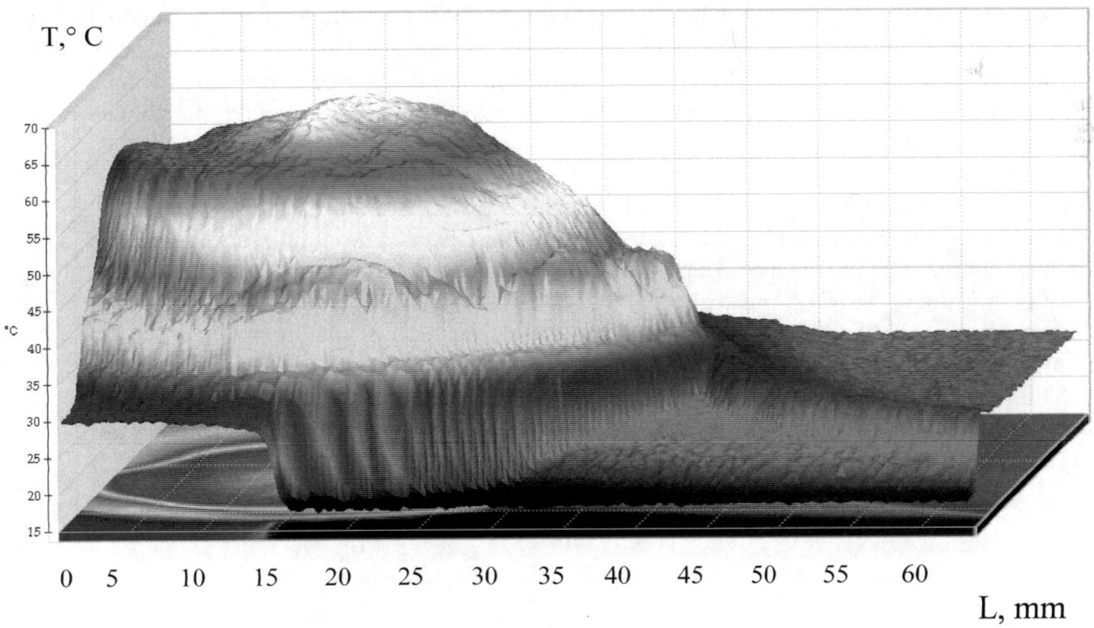

Figure 1. 3-D thermogram of magnetically controlled TM (thermal accumulating material).

IOP Conf. Series: Journal of Physics: Conf. Series **1084** (2018) 012013 doi:10.1088/1742-6596/1084/1/012013

Figure 2 shows the thermogram of magnetically controlled TM in the mode of thermomagnetic convection.

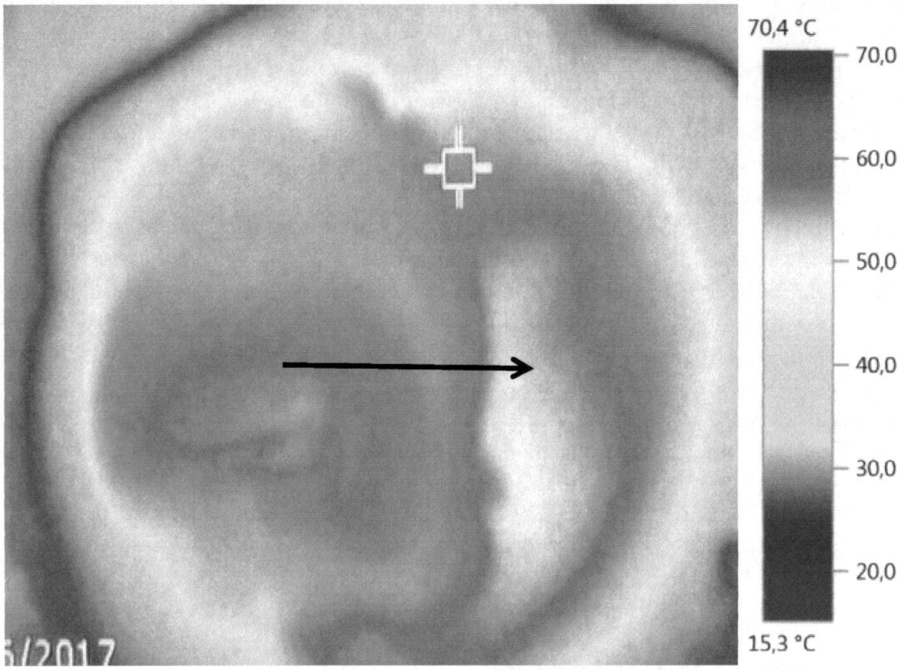

Figure 2. Thermogram of magnetically controlled TM (thermal accumulating material).

As the magnetically controlled TM is heated the magnetic properties change; and the cooler layer displaces the hot layer. Thus, thermoregulation on the heater surface and interaction with the environment is achieved. Since the developed TM can provide thermoregulation due to the effect of thermomagnetic convection and intensification of heat transfer does not require additional energy, this approach is economically expedient and technologically adapted for aerospace engineering.

4. Methods for obtaining analytical functions for the thermoregulation system

The efficiency of the thermoregulation system with TA (thermal accumulators) is related to the efficiency of optimizing the constructive parameters of the heater. It is possible to achieve this through mathematical modeling based on finite integral transform method.

The main question of mathematical modeling is the distribution of the temperature field of TA in the regime of thermomagnetic convection in various spatial volumes.

Modeling of TA is carried out taking into account that its form can be represented by unlimited multilayered plate, unlimited continuous and hollow multilayered cylinder, solid and hollow multilayered ball [9] Figure 3.

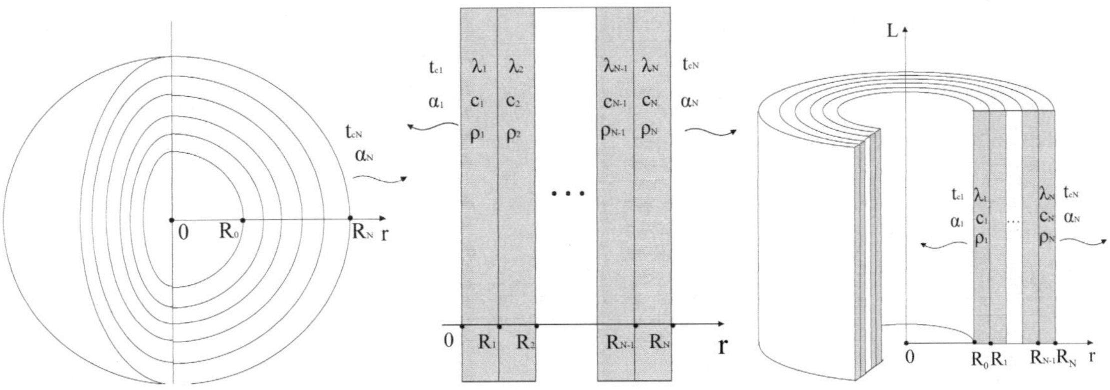

Figure 3. Multilayered bodies of canonical form: unlimited plate, ball, unlimited cylinder.

Consider mathematical model of TA with random initial conditions, inhomogeneous boundary conditions and distributed internal heat source.

$$\frac{\partial t_i(r_i,\tau)}{\partial \tau} = a_i^2 \left(\frac{\partial^2 t_i(r_i,\tau)}{\partial r_i^2} + A_{k,i} \frac{\partial t_i(r_i,\tau)}{\partial r_i} \right) + Q_i(r_i,\tau);$$

$$i = 1,2,\ldots,N; \qquad R_{i-1} \le r_i \le R_i; \quad k = 0,1,2; \qquad \tau > 0; \tag{1}$$

$$t_i(r_i,0) = f_i(r_i); \tag{2}$$

$$\lambda_1 \frac{\partial t_1(R_0,\tau)}{\partial r_1} + \alpha_1 \left(t_1(R_0,\tau) - t_{c1}(\tau) \right) = 0; \qquad \alpha_1 < 0; \tag{3}$$

$$\lambda_N \frac{\partial t_N(R_N,\tau)}{\partial r_N} + \alpha_N \left(t_N(R_N,\tau) - t_{cN}(\tau) \right) = 0. \tag{4}$$

$$t_j(R_j,\tau) = t_{j+1}(R_j,\tau); \quad \lambda_j \frac{\partial t_j(R_j,\tau)}{\partial r_j} = \lambda_{j+1} \frac{\partial t_{j+1}(R_j,\tau)}{\partial r_{j+1}}, \tag{5}$$

$$j = 1,2,\ldots,N-1.$$

Here:

r – space coordinate;

τ –time;

N – number of layers in a multilayered area;

$t_i(r_i,\tau)$ – temperature field of the i-area;

$Q_i(r_i,\tau)$ – function of the internal heat source of the i-area;

λ_i, α_i^2 – coefficients of thermal conductivity and thermal diffusivity of the i-area, respectively;

$A_{k,i}$ – the equation coefficients determined by the type of coordinates:

– for a Cartesian coordinate system k=0, $A_{0,i} = 0$;

– for a cylindrical coordinate system k=1, $A_{1,i} = 1/r_i$;

– for a spherical coordinate system k=2, $A_{2,i} = 2/r_i$;

α_1, α_N – coefficients of convective heat transfer from external surfaces to the environment;

$t_{c1}(\tau)$, $t_{cN}(\tau)$ – environment temperatures as the functions of time;

R_{i-1}, R_i – coordinates of the i-area boundaries.

To exclude the coordinate r, along which the body properties change stepwise, the formula for the transition to images is used:

$$U(\mu,\tau)=\sum_{m=1}^{N}\frac{\lambda_m}{a_m^2}\int_0^{R_m}\rho(z_m)t_m(z_m,\tau)W_m(z_m,\mu)\,dz_m, \tag{6}$$

where $\rho(z)$ – is a weight function that is a solution of equation

$$\frac{d\,\rho(z_m)}{d\,z}-A_{k,m}\;\rho(z_m)\;=0. \tag{7}$$

The kernel of the integral transformation $W_m(z_m,\mu)$ is a solution of the Sturm-Liouville problem with the corresponding homogeneous boundary conditions, determined to within constant factor (here - parameter):

$$\frac{d^2W_m(r_m,\mu)}{d\,r_m^2}+A_{k,m}\frac{d\,W_m(r_m,\mu)}{d\,r_m}+\frac{\mu^2}{a_m^2}W_m(r_m,\mu)=0;$$
$$m=1,2,\ldots,N,\qquad R_{m-1}\le r_m\le R_m; \tag{8}$$

$$\lambda_1\frac{d\,W_1(R_0,\mu)}{d\,r_1}+\alpha_1\,W_1(R_0,\mu)=0; \tag{9}$$

$$\lambda_N\frac{d\,W_N(R_N,\mu)}{d\,r_N}+\alpha_N\,W_N(R_N,\mu)=0; \tag{10}$$

$$W_j(R_j,\mu)=W_{j+1}(R_j,\mu);\quad \lambda_j\frac{d\,W_j(R_j,\mu)}{d\,r_j}=\lambda_{j+1}\frac{d\,W_{j+1}(R_j,\mu)}{d\,r_{j+1}},$$
$$j=1,2,\ldots,N-1. \tag{11}$$

The solution of equation (8) has the form:

$$W_m(r_m,\mu)=C1_m\,Exp\left(-\frac{r_m}{2}\left(A_{k,m}+\sqrt{A_{k,m}^2-4\frac{\mu^2}{a_m^2}}\right)\right)+$$
$$+C2_m\,Exp\left(-\frac{r_m}{2}\left(A_{k,m}-\sqrt{A_{k,m}^2-4\frac{\mu^2}{a_m^2}}\right)\right). \tag{12}$$

In the Cartesian coordinate system

$$W_m(r_m,\mu)=C1_m\,Sin\left(\frac{\mu}{a_m}r_m\right)+C2_m\,Cos\left(\frac{\mu}{a_m}r_m\right). \tag{13}$$

In the cylindrical coordinate system

$$W_m(r_m,\mu)=C1_m\,J_0\left(\frac{\mu}{a_m}r_m\right)+C2_m\,Y_0\left(\frac{\mu}{a_m}r_m\right), \tag{14}$$

where $J_0(z)$, $Y_0(z)$ – Bessel functions of the first kind, zero and first order, respectively.

5. Conclusion

There were investigated the mechanisms of thermomagnetic convection in the thermal accumulating material based on the synthetic oil containing ferromagnetic particles (nickel-zinc ferrite), paraffin and MWCNTs under the influence of magnetic field created by neodymium magnet with magnetic induction of 1000 mT. It was found that during the heating of the magnetically controlled TA the magnetic properties change; and the cooler layer displaces the hot layer.

The mathematical modeling of TA is carried out taking into account the fact that its shape can be represented by unlimited multilayer plate, unlimited continuous and hollow multilayer cylinders, solid and hollow multilayer balls. Analytical equations are obtained for determining the temperature field in the volumes containing TA.

The developed magnetically controlled thermal accumulator can provide thermoregulation due to the effect of thermomagnetic convection and intensification of heat transfer does not require additional energy. Such approach is economically expedient and technologically adapted for aerospace engineering.

Acknowledgments

This paper was financially supported by the Ministry of Education and Science of the Russian Federation on the project No.26.1146.2017/4.6 «Development of mathematical methods to forecast efficiency of using space services in the national economy».

References

[1] Liping Pang, Yue Qin, Dong Liu and Meng Liu 2014 *Chinese Journal of Aeronautics* **27(2)** 210-216

[2] Strøm-Tejsen P, Zukowska D, Jama A and Wyon D P 2007 *Proceedings of Roomvent* **1** 227–234

[3] Shchegolkov A, Shchegolkov A, Dyachkova T, Bodin N and Semenov A 2017 *AIP Conference Proceedings* **1899** 020020

[4] Renteria J, Legedza S, Salgado R, Balandin M, Ramirez S, Saadah M, Kargar F and Balandin A 2015 *Materials and Design* **88** 214-221

[5] Correa-Duarte M A, Grzelczak M, Salgueiriño-Maceira V, Giersig M, LizMarzán L M, Farle M, Sierazdki K and Diaz R 2005 *J. Phys. Chem. B* **109 (41)** 19060–19063

[6] Garmestani H, Al-Haik M, Dahmen K, Tannenbaum R, Li D, Sablin S and Hussaini M 2003 *Adv. Mater.* **15(22)** 1918–1921

[7] Yuan F, Jiao W, Yang F, Liu W, Xua Z and Wang R 2017 *RSC Adv.* **7** 43380–43389

[8] Salehirad M, Alavi Nikje M M and Ahmadian-Alam L 2018 *Industrial & Engineering Chemistry Research* **57(6)** 1803-1814

[9] Shchegolkov A, Borovskikh P and Shirshova L 2017 *MATEC Web of Conferences* **129** 01050

IOP Conf. Series: Journal of Physics: Conf. Series **1084** (2018) 012014 doi:10.1088/1742-6596/1084/1/012014

Estimation and increase of information measuring systems' metrological reliability on the designing stage

T Chernishova[1], V Tretiakov[1] and N Othman[2]

[1]Institute of Energy, Instrument Making and Radio Electronics, Tambov State Technical University, 106 Sovetskaya Street, Tambov 392000, Russian Federation
[2]Department of Engineering and Information Technology, Al Azhar University - Gaza, Almadaris street House 6, Gaza Strip, Alburige, Palestine

[1]E-mail: energo@nnn.tstu.ru

Abstract. The paper deals with mathematical modelling application aimed at the estimation and increase of the metrological reliability of the information measuring systems on designing stage. The algorithm of metrological reliability estimation is based on mathematical modeling of non-stationary random processes of temporal variation of the metrological characteristics of the measuring means. There is a discussion of two ways to realize the developed method of increasing metrological reliability of the information measuring systems in the design phase based on mathematical modeling and use of parametric optimization methods. The obtained results make it possible for customer to develop information measuring systems with counted value of metrological reliability and also to form the design decisions in order to make information measuring system with maximum value of metrological resource.

1. Introduction

Nowadays realization of any modern project in the industrial design aria is completely impossible without measuring tools application. Currently, among the different groups of measuring devices the use of information-measuring systems (IMSs) is increasing [1]. IMSs are used in any experimental research or manufacturing enterprise. Precision of the got experimental data or quality of the output goods depend on metrological reliability of the used IMS [1, 2]. So, IMS' complexity and responsibility set the very actual problem to provide high metrological reliability of the used IMSs.

As shown by theoretical and practical researches the most important parts of the measuring devices which are responsible for metrological reliability are generally analog blocks as sensors, amplifiers, converters, etc. [2], which transform analog signals of measured physical values before its conversion to the digital form. Meanwhile, metrological reliability is a basic characteristic of IMS' quality. Herewith, the main quantitative indicator of metrological reliability is metrological resource determined by the time when the measuring inaccuracy surpasses the acceptable limits [3].

So, the aim of our research was to develop the methods providing estimation and increase of IMS' metrological reliability on the designing stage with mathematical modeling application. Realization of these methods lets to avoid big financial expenditures and complication of the IMS' structure.

2. Experimental part

2.1. Estimation of IMS metrological reliability

The metrological reliability of IMS analog blocks was estimated by analytic-probabilistic forecasting based on the mathematical modeling of non-stationary random processes time-depended metrological characteristics of measuring instruments. The main metrological characteristic of an instrument was measuring inaccuracy δ.

The first step of the algorithm included simulation of the circuitry of IMS analog blocks by the equation (1):

$$\xi_i = F_1\left(\xi_{i0}, \vec{\phi}, t\right) \tag{1}$$

where $\vec{\phi} = \{\phi_1, \phi_2, ..., \phi_m\}$ was vector of environmental factors, m – number of environmental factors, $\xi_i(\vec{\varphi}, t)$ – parameter value of the component number i, t – exploitation time, and ξ_{i0} – nominal value of the component number i.

Then we developed the mathematical model of functioning analog block. The mathematical model introduced the relation between output y and input x signals, parameters of analog block components $\vec{\xi} = \{\xi_1, \xi_2, ..., \xi_n\}$ and environmental factors $\vec{\phi} = \{\phi_1, \phi_2, ..., \phi_m\}$ in the equation (2):

$$y = F_2\left(x, \vec{\xi}, \vec{\phi}\right). \tag{2}$$

The mathematical model (2) was developed on the base of electric scheme of the block and electric schemes calculation methods. It was necessary to use mathematic formula of explored metrological characteristic S to research block metrological properties. Therefore, the model (2) transformed to the model (3):

$$S = F_3\left(\vec{\xi}, \vec{\phi}, t\right). \tag{3}$$

Electronic components of IMS analog blocks as capacitors, resistors, chips, etc., have a tendency to aging, it brings deviation of components parameters from the nominal values. It garbled the measuring signal and as a result increased a measuring inaccuracy δ.

The environmental factors (e.g., temperature, humidity, pressure, radiation) effect expedites the degradation processes which progress in the components of IMS analog blocks. It decreases the metrological source as a main parameter of the metrological reliability.

The summarizing mathematical model of the metrological characteristic of analog blocks of IMS (that is an inaccuracy δ) gets the view:

$$\delta(t) = F_4\left\{x, \vec{\xi}(t), \vec{\phi}(t)\right\}. \tag{4}$$

The developed mathematical model (4) took into account the analog blocks components parameters variation under environmental factors impact. This model was used in the procedure of statistical modeling of metrological characteristic at different time of exploitation. The modeling algorithm consisted of consecutive characteristics calculation of the distribution mechanism of the components parameters values of the designing analog blocks and metrological characteristics modeling at the different time moments.

The mathematical model describing variation of time-depended metrological characteristics was translated into analytical complex equations of mathematical expectations $m_\delta(t)$ variation in time functions and functions describing the variation in time of possible values borders for metrological characteristic deviations from its mathematical expectation $\psi_{\pm\sigma}(t)$ was built on the base of the statistical modeling results:

$$\begin{cases} m_\delta\left(\vec{\phi},t\right), \\ \psi_{\pm\sigma}\left(\vec{\phi},t\right)=m_\delta\left(\vec{\phi},t\right)+\mathrm{c}\cdot\sigma_\delta\left(\vec{\phi},t\right) \end{cases} \tag{5}$$

where c was a coefficient which depends on the level of the trust probability P, c = 3 if P = 0.9973.

The developed mathematical model (5) was the basis for calculation of metrological resource values by extrapolation of the equations (5) on future time in exploitation area. Exactness of the prediction results depended on the exactness of the using mathematical models (1) for analog blocks components and also on adequacy of built mathematical models of metrological characteristics.

2.2. Increase of IMS metrological reliability

IMS metrological reliability was increased through the growth of the metrological resource via the algorithm presented on the Figure 1.

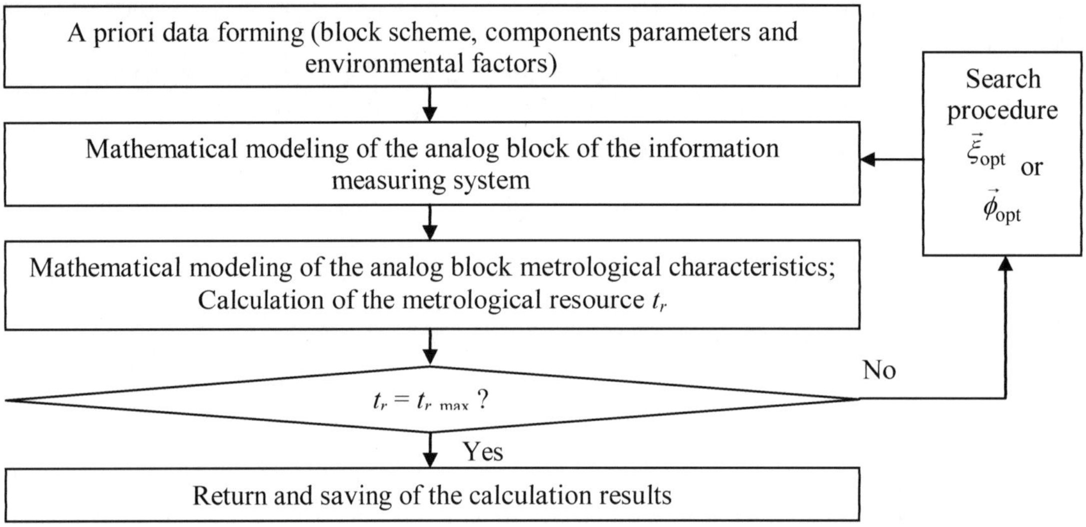

Figure 1. The algorithm of IMS metrological reliability increase.

There were two possible ways to increase the metrological reliability of IMS analog blocks:

– optimal choice of nominal components providing the highest level of metrological resource under static environmental conditions, and

– optimal choice of exploitation conditions (environmental factors $\vec{\phi}=\{\phi_1,\phi_2,\ldots,\phi_m\}$) for a basic components and general analog blocks giving the highest level of the metrological resource of the developed analog blocks.

The mathematical form of this task was:

$$t_r^* = \max\left\{\min_{l=1,\ldots,L}\left\{t_{rl\_\max}\right\}\right\} \text{ if } \delta_l\left(t,\vec{\xi},\vec{\phi}\right)=\text{var}, \\ \vec{\phi}=M, y_l\left(t,\vec{\xi},\vec{\phi}\right)\in A_l \tag{6}$$

where $y_l\left(t,\vec{\xi},\vec{\phi}\right)$ – output characteristic of analog block number l, l = 1, 2, …, L – number of analog blocks in the information-measuring system; $\delta_l\left(t,\vec{\xi},\vec{\phi}\right)$ – measuring inaccuracy of analog block number l; $\vec{\phi}$ – vector of environmental factors; M –range of possible environmental factors values; $\vec{\xi}$

–vector of components parameters values; A_l –diapason of correct output signal values of analog block number l; t –exploitation time moment; t_r^* –metrological resource of information-measuring system.

The most appropriate way to solve the task described by the formula (6) was the parametrical optimization method since it provided the max value of the metrological resource by modeling components parameters values and exploitation conditions,.

Statistical modeling of the accuracy δ for the researching analog blocks included analog block parameters components modeling based on change time-depended values and environmental factors impact and modeling of the researching metrological characteristic realization at different time $\delta(t_k)$, $k = 0, 1, \ldots, K$ (K – the number of the time points which the metrological characteristics).

The last step of the algorithm is developing the mathematical model of the change in time process of the metrological characteristic (5) for each analog block is developed. Then the original metrological resource value of the researching analog block is defined by extrapolation of parameters $m_\delta(t)$ and $\psi_{\pm\sigma}(t)$ on the area of the future exploitation.

2.3. Approbation of the developed algorithms

The developed models were realized and helped to find optimal values of component parameters $\vec{\xi}_{\text{opt}}$ and environmental factors $\vec{\phi}_{\text{opt}}$ for each IMS analog block. Firstly, we defined the most important components of the block parameters and vital environmental factors. Then, the selected parameters were processed by the method of configurations (Hooke-Jeeves method) [4]. The research procedure of the maximum value of the metrological resource was realized using the searching method of configurations.

The important part of the algorithm is control of the functionality condition of the analog block on each step of the algorithm:

$$y\left(t,\vec{\xi},\vec{\phi}\right) \in A \qquad (7)$$

the $y\left(t,\vec{\xi},\vec{\phi}\right)$ output characteristic of the analog block, A is the diapason of the correct output signal values of the analog block.

For the approbation we used voltage frequency converter as typical analog block which enters into the composition of various IMSs. Here we have chosen environmental factors having impact on block metrological resource as temperature T, humidity F, pressure P and radiation effect E and the initial values of these factors: T_0 = 20 °C, F_0 = 45 %, P_0 = 1 atm, E_0 = 50 μR/h. Also there were defined the most valuable of these factors: temperature T and humidity F.

Then the value of metrological resource corresponding to initial factors was calculated as 40500 hours by equation (2). The detailed calculations were missed in this article because it introduced in the article [5]. Time-dependent inaccuracy variation before and after developed method application is shown on the Figure 2.

We found the optimal values of the chosen most valid factors: optimal temperature T_{opt}=15 °C and humidity F_{opt}=25%. These environmental factors values provide the maximal value of the metrological resource of the voltage-frequency converter: 46800 hours. The maximum inaccuracy value is 5 % (Figure 2).

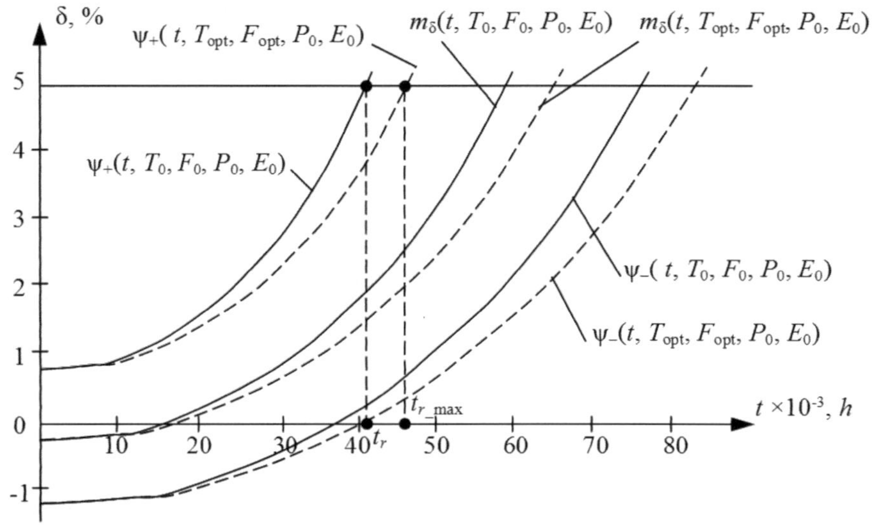

Figure 2. The calculated metrological reliability of block parameters and environmental factors: δ – inaccuracy value, t_r – metrological resource value, t_{r_max} – metrological resource maximum value, t – time, h – hours, T_0 – initial value of temperature, F_0 – initial value of humidity, P_0 – initial value of pressure, E_0 – initial value of radiation, T_{opt} – optimal value of temperature, F_{opt} – optimal value of humidity, m_δ – mathematical expectations, ψ_-, ψ_+ – metrological characteristic deviations from its mathematical expectation .

Conclusion

The searching algorithm of the optimal values of the components parameters and the algorithm of the environmental condition optimization which provide the maximum level of metrological resource are developed in this article

We can make a conclusion that realization of the offered algorithms of the metrological reliability increase of the analog blocks of the IMS which includes mathematical modeling on the designing stage of the analog blocks makes it possible to increase the metrological resource as a main characteristic of the metrological reliability more than 10 %.

The obtained results make it possible for customer to develop IMSs with counted value of metrological reliability and also to form the design decisions in order to make information-measuring system with maximum value of metrological resource. What is more, realization of these methods lets to avoid big financial expenditures and complication of the IMS' structure.

References

[1] Stakhov A 2016 Fibonacci p-codes and Codes of the "Golden" p-proportions: new informational and arithmetical foundations of computer science and digital metrology for mission-critical applications *British Journal of Mathematics & Computer Science* **17(1)** 1

[2] Monteiro C and Mari E 2015 *XXI meko World Congress "Measurement in Research and Industry"* Prague, Czech Republic, 30 August – 4 September 2015 **1**

[3] Bucci D 2017 Analog electronics for measuring systems *Focus Electronics Engineering Series* (Great Britain: John Wiley & Sons, Inc)

[4] Yaman K and Arslan G 2017 The optimization of solar water heating system using hybrid algorithm (PSO/HJ) for different locations of Turkey *Int. J. Eng. Res. Develop.* **9** 3

[5] Chernyshova T I and Tretiakov V V 2017 Method of the metrological reliability increase in disigning of the analog parts of the information-measuring systems *Reliability and Quality of the Complex Systems* [in Russian] **17(1)** 50

Institute of Physics
Dirac House, Temple Back
Bristol BS1 6BE UK

ISSN: 1742-6588
ISBN 978-1-5108-7512-8

9 781510 875128

CURRAN ASSOCIATES INC.
proceedings.com